Visualizing the Structure of Science

Benjamín Vargas-Quesada · Félix de Moya-Anegón

Visualizing the Structure of Science

With 81 Figures and 53 Tables

 Springer

Authors

Benjamín Vargas-Quesada

Faculty of Information and Library Science
University of Granada
Colegio Máximo de Cartuja s/n
18071 Granada
Spain

benjamin@ugr.es

Félix de Moya-Anegón

Faculty of Information and Library Science
University of Granada
Colegio Máximo de Cartuja s/n
18071 Granada
Spain

felix@ugr.es

The scientograms can be seen in color on the book's page at springer.com/978-3-540-69727-5

Library of Congress Control Number: 2007920900

ACM Computing Classification (1998): H.5, H.4, J.4

ISBN 978-3-540-69727-5 Springer Berlin Heidelberg New York

Springer is a part of Springer Science+Business Media
springer.com
© Springer-Verlag Berlin Heidelberg 2007

Typesetting: by the authors
Production: Integra Software Services Pvt. Ltd., Pondicherry, India
Cover design: KünkelLopka, Heidelberg

Printed on acid-free paper   45/3100/Integra   5 4 3 2 1 0

Table of Contents

1 Introduction

The construction of a great map of the sciences is a persistent idea of the modern ages. This need arises from the general conviction that an image or graphic representation of a domain favors and facilitates its comprehension and analysis, regardless of who is on the receiving end of the depiction, whether a newcomer or an expert.

Researchers are well aware of the limitations existing for the generation of new knowledge from a single discipline. For this reason, they have long combined the resources, techniques and methodologies of different subject matters to foster the appearance of new knowledge, and even of new disciplines. Biotechnology is just one example. The "globality of science", the fruit of informational fluxes among disciplines, makes the underlying structure of the scientific world increasingly complex and difficult to analyze.

Sociological resources such as social network theory grant us one particular means of entry into the study of science, in view of the relationships of its component parts. Graphs known as sociograms are able to depict hidden social networks, while graph theory, which provides the conceptual foundation for structural analyses, makes it possible to visualize the relations existing among disciplines and interpret them.

Yet even social networks have a tendency to overflow, with their vast and moving volume of structural complexities, due to the increase of communication across disciplines over the years.

Browsing through the literature on social networks and information visualization, we came across PathFinder NETworks (PFNET). This is an algorithm that prunes network links, in such a way that a graph is left clean of superfluous relationships, and maintains only the salient ones, thereby favoring visualization and the analysis of contents.

This process of simplifying the panorama of science is not really new or unique to our days. It evokes a principle of reason formulated at the end of the Middle Ages, known to many and under various denominations: the principle of economy, the principle of parsimony or simplicity, or even "Ockhams's razor", because of its attribution to the Franciscan philosopher William of Ockham (though in fact it can be traced back to times even before then).

In its original fifteenth century formulation, in Latin, this principle of simplicity states that *pluralitas non est ponenda sine necessitate*, that is, that the essential things should not be multiplied unless necessary. In a more colloquial language, we could say that we should not create too much of a good thing; or that the hypothesis behind a line of reasoning ought to be left in its simplest terms, as reasoning based on less numerous premises will ring truer.

In the field of science, we say that one should favor the most simply stated hypothesis that explains (or attains to explain) one's observations. This principle has been carried much further since the scientific community began to admit that the laws of physics preferably be written in the language of mathematics: the best hypothesis is, a priori, the one that has the simplest mathematical formulation.

The visualization and analysis of scientific structures is welcome terrain for this very minimalistic principle: why visualize a dense and complex structure if, instead, we can accurately derive another one that is simpler to interpret, harboring all the clearly significant or essential relationships?

2 Visualization

Scientific information is spread out over disciplines which, to the outside observer, may seem to have little in common. For this reason, when traditional methods are used to study a domain pertaining to one specific field of knowledge, one is sometimes left with a sensation of not grasping the domain as a whole. It is like trying to complete a puzzle and not knowing where to put the piece held in the hand, not seeing which puzzle pieces it fits in with.

The representation of scientific information in ways easier for the human mind to embrace is nothing new. "To make visible to the mind that which is not visible to the eye, or to create a mental image of something that is not obvious (e.g. an abstraction)" (Owen, 1999) is the definition of the word "visualization" that point to the intrinsic need to represent information in a non-traditional manner. To paraphrase *Costa* (1998), visualizing is neither the implicit result of the act of seeing nor a spontaneous product of the individual receiving visible input. To visualize is a task of the communicative process, through which abstract data and complex phenomena of reality are transformed into visible messages. This enables individuals to apprehend with their own eyes certain data and phenomena that cannot be directly retrieved from a hidden body of knowledge.

Although relatively young, the field of computer visualization has expanded in a number of directions in a very short time. It is used to make visible to the human eye that which is small and very difficult to perceive, such as the molecular structure; or too vast to be comprehended, for instance imagery of our universe. In its quest to visually display that which only exists in our minds, information science resorts to the diffuse realm of the so-called virtual reality. Computerized visualization techniques can be used to make manifest phenomena that are not visible on their own, such as the implicit relationships among component elements (Araya, 2003).

2.1 Visualization Technology

From the study by *Araya* (2003), and under the auspices of the development of information science, along with other disciplines, computerized visualization took a giant leap in the mid-1980s after a report by the

National Science Foundation (NSF), in which a number of problems were addressed in the context of the scientific community. The main one is that of the dilemma of uninterpreted information. This refers to the diversity of sources such as satellites, medical scanners, radars, and so on, and the complexity of the information that they supply, for which processing and interpreting the results is tremendously difficult. Other key matters are how to enhance the communication of results among scientists, and how to achieve greater interaction between researchers and the computerized analysis of data. The solution to these problems evokes the development of visualization technology:

> Scientists need an alternative to numbers. A technical reality today and a cognitive imperative tomorrow is the use of images. The ability of scientists to visualize complex computations and simulations is absolutely essential to insure the integrity of analyses, to provoke insights and to communicate those insights with others. (McCormick; DeFanti; and Brown, 1987)

2.2 Information Visualization

The graphic representation of information for its posterior visualization is an activity common to most scientific disciplines of recent times (Klovdhal, 1981) (Crosby, 1997). But the use of graphic representations in conjunction with computer technology to attain an adequate visualization of the information is a relatively new task, which has rapidly become one of the main objectives of research in the 1990s.

The visualization of information, as *Costa* explains, is the undertaking of a visual communicator, which transforms abstract data and complex phenonomena of reality into visible messages, thus enabling individuals to see with their own eyes the data and phenomena that are directly unapprehensible and conform a body of hidden information (Costa, 1998). The domain of graphic languages for visualizing these invisible effects or unapprehensible phenomena configure a new science of visual communication – *Schematics* – which *Costa* defines as *the third language*, after the image and the sign.

Meanwhile, the visualization of information also requires the transfer of knowledge. Visualization can be justified by the fact that the world is multi-faceted, multidimensional, multiphenomenal, and is presented as a continuum. Therefore, to visualize information we need to arrive at an intersection of image, word, number, and art. The instruments needed to reach this cross-roads are related with writing and typography, the management and ensuing statistical analysis of vast amounts of data, and graphics, their distribution, and color. Essential standards of quality are also derived from visual principles, which provide solid clues as to what to display, and where (Tufte,

1994). A major difficulty arises at the time of transferring the complexity, dynamism, and multidimensionality of any component of our world to a compressed, reduced, static, and flat format, such as paper or the computer screen. To overcome these obstacles, *Tufte* proposes six ideas:

1. Avoid, at all times, flat representations and use three-dimensional graphics, or at least two-dimensional ones with perspective (2.5D).
2. Construct micro- and macrorepresentations of the same single reality; avoid disorderly groupings of elements, instead shedding light on details and complex elements by means of the effective distribution of information and its stratified visualization.
3. Capture different images of a single reality from different perspectives.
4. Associate colors with information.
5. Achieve graphs in four dimensions, which entail the combination of tridimensional graphs with related temporal information.

By fusing the ideas of *Costa* and *Tufte*, we can say that visualization is a process of communication between a reduced representation of reality and the person who observes it, and from which it is possible to perceive through eyesight certain facts and phenomena of multidimensional and ever-changing world reality that would otherwise slip by unnoticed. Further, we can state that the visualization of information resides at the intersection of image, word, number and art; and takes form through writing and typography, the processing of vast amounts of data and statistical analysis, graphics, layout and color. All this for the end purpose of obtaining a reduced graphic representation of a multidimensional and ever-changing reality, relaying to the observer the facts and phenomena of a certain portion of reality which, without the participation of all the above, would remain beyond our conscious awareness.

The visualization of information is by no means a new practice in the field of Documentation: suggested over 60 years ago by *Bush* (1945), and put into practice just over 40 years ago by *Garfield, Sher and Torpie* (1964), the visualization of scientific information has long been used to "uncover" and divulge the essence and structure of science. Yet despite its ripe age, information display is still in an adolescent stage of evolution in the context of its application to scientific domain analysis, among other promising realms of development.

3 Visualization of Scientific Information and Domain Analysis

The analysis of domains is one of the newest research fronts to surge forward as a consequence of the proliferation of information visualization techniques, and proves immensely helpful in revealing the essence of scientific knowledge (Chen; Paul; and O'keefe, 2001). For example, it has been used to show animated visualizations about the extinction of mass media literature, or to demonstrate the potential of visualization based on citation (Chen [et al.], 2002), as well as to explore and access the contents of digital libraries (Chen, 1999b) and to study the evolution of the patterns of citation regarding patents (Chen and Hicks, 2004).

Of course, there is a close but perhaps elusive connection between domain visualization and that which *Hjørland and Albrechtsen* (1995) call domain analysis. The visualization of domains can afford a strong means of support in domain analysis, especially within areas of multidisciplinary knowledge, and in those fields that change and advance rapidly (Börner; Chen; and Boyack, 2003).

Although the relationship between the visualization and the analysis of domains, or visualization of knowledge, has been suggested by authors *Garfield*, *Small*, *White* and *Chen* among others, none has come to explain the whys and whereabouts of this deeply rooted association. What tools, or technical aids, can make a contribution to domain visualization and domain analysis, and vice versa. We believe it is high time to address this question, with a study that adopts a singular perspective or thesis: the analysis of domains based on their visualization, through representation via social networks.

3.1 Domain Analysis

In 1995, *Hjørland* and *Albrechtsen* set forth a new approach or perspective for studying the field of Documentation and Information Science: domain analysis. According to this new viewpoint, the analysis of domains is based on the analytical domain paradigm, which establishes that the best way to understand information is to study a given domain of knowledge as part of the discourse of the communities from which it proceeds, the community being a precise reflection of economic and work divisions of society. This is due to the fact that the organization of knowledge, its structure, patterns of cooperation, language and modes of communication, and criteria of relevancy are the mirror image of the work of these communities and of the role that they play in society. Moreover, the individual psychology of each member, his or her knowledge, the informative need of a person, and the subjective criteria of relevance of each are taken into account. The domain-analytical paradigm is, in the first place, a social paradigm in that it foments a psychological perspective, and sociolinguistic and sociological perspectives of science. Secondly, it is a functionalistic approach, as it attempts to comprehend the implicit as well as the explicit aspects of science while marking or making visible underlying mechanisms of communication. And in the third and final place, it provides for a philosophical–realistic approach, as it aims to establish the scientific bases of a given domain, through factors that remain external to individual perceptions and the subjective realm of users – in contrast, for instance, to the cognitive and conductivistic paradigms (Hjørland and Albrechtsen, 1995).

This perspective is not necessarily limited to of the area of Documentation, but can be applied to any other domain, regardless of its nature or size. In our opinion, what is important is not the development of a new theory or paradigm to let us analyze each one of the disciplines now in existence; rather, the paradigm should ideally form a conceptual baseline for developments in any field, even in the earliest stages of development. Be it separately or in combination, as a set, this can be achieved through a holistic and objective view of such domains.

Whether in Philosophy or in scientific theory, then, there is a trend from the fundamentalistic or empiric theories asserting that science is based on postulates of absolute truths obtained by means of the human senses – empirical thought and positivism – and from reason – rationalism. This positivistic and rationalistic view of science considers language to be something nominalistic, its only utility being that of transporting perceived knowledge – acquired by the senses or by reason. Accordingly, language does not intervene in the process of the perception of reality, but is simply the vehicle through which already existent knowledge is communicated among individuals. This focus emphasizes the individual perception of knowledge, free from cultural traditions; yet in light of its overly objective view of things, it is a philosophy that is outdated. And so, the traditional perspective of epistemology and the Philosophy of Science is being replaced by a more holistic current that acknowledges the importance of language in the perception of reality and which is consequently introducing a historical, cultural, social, and objective dimension into the theory of knowledge and of science. Reality cannot be captured nor be comprehended innocently by an isolated and unprepared individual. Even individualized bodies of knowledge are constructed with elements of history, culture and whatnot, including building materials that may have developed from a more specific domain of knowledge offering one a slightly distinctive possibility of perceiving reality. In short, the methodological individualism that considers knowledge as a mental state of the person is being phased out by a methodological collectivism or holism, which grasps knowledge as a social or cultural process, or as a cultural product (Hjørland and Albrechtsen, 1995).

From this standpoint, we might say that the best way to study a discipline or domain of knowledge would be through *domain analysis*, understood as an illuminating study of the discourse of the community participating in the formation or evolution of a discipline, as well as of certain hidden relationships with the society in which it gestates. In this way, the social, ecological and informative nature of the domain of knowledge under study is accented.

But now, what tools, methods, techniques, and so forth are necessary in order to carry out the analysis of a domain? *Hjørland* again comes up with a response, proposing 11 methods for the analysis of domains within Documentation (Hjørland, 2002). We would add that these methods (not exclusive to Documentation) may be used to discover a holistic and objective panorama within any sort of domain.

Of the 11 methods proposed by *Hjørland* for domain analysis, we focus on one, specifically that of bibliometric studies, as constituting the best approach whence to ponder domain analysis. The other ten may well be used as complementary elements to back-up and give more substance to this holistic vision.

Until fairly recently, sociologists believed that bibliographic citations were some sort of system for the control of intellectual property safeguarded in scientific publications. The importance that they wielded, additionally, in reflecting cognitive and social connections among researchers went unacknowledged (Merton, 2000). But in the field of Documentation, authors soon began to appreciate this alternative facet of citation. Networks borne through the citation of scientific documents can clearly signal the emergence of new research fronts (Price, 1965) just as they can be used to obtain ethnographic information referring to the presence and nature of social relations – for example, to discover through citation a close colleague whom one has never met in person (White, 2001). The use of this technique can be extended beyond bibliometrics or sociology to become a general notion in which different subdisciplines flow together, including: scientometrics, infometrics, and bibliometrics in the strict sense. Because of the focus of our research, for us it has become a synonym for metric studies surrounding science.

Bibliometrics configure a discipline with a long tradition, which has had to adopt different techniques with the passing of time. In its beginnings it involved mathematical distributions, the creation of rankings, with bivariate analysis; in later years, it has introduced a wide array of techniques involving Multivariate Analysis (MA). Among these we find statistical methods, connectionist techniques, and social networks, which we will deal with in greater detail under Sect. 4.4.

Cocitation analysis may also be used by Documentation professionals who wish to find out how one given author is related with the rest of the scientific community in a specific or a broad area of work; by those who wish to study the structure of a scientific domain; or by those who want to gain awareness of new advancements on the horizons of science. It would be very desirable in general to be able to combine citation analysis with social network analysis in order to explore how social structures penetrate or are reflected by the intellectual structure of the individuals therein. But until now, bibliometric experts did not dispose of tools geared to the study of social networks, and sociologists were not tuned into the utility of citation as an informational source on its own accord nor of the developments in visualization techniques that might be applied to this end (White, 2000).

Hjørland defends bibliometrics as a tool and method for domain analysis in many different forms. For example, as a tool it can be used for the generation of bibliometric maps by means of cocitation analysis. This is the case of the maps for the visualization of a discipline (White and McCain, 1998b), whose representations bring to light factors that are external to the user's subjective perception, by breaking through aprioristic mental schemes and, as a consequence, facing the representation of a reality that was not previously perceived. Moreover, bibliometrics can be used as a method for the analysis of domains, as it shows the real relationships between and among individual documents and reveals the explicit acknowledgment that some authors make of others, while in the meantime reflecting relationships among different scientific fields (Garfield, 1976).

Bibilometrics can show and describe tendencies in different areas of knowledge, but in itself it cannot interpret the utility, the fit, the benefits, and the drawbacks of such tendencies. For this purpose we need to resort to broader disciplines such as Sociology and Philosophy, which permit a sociocultural interpretation of data and of bibliometric displays (Hjørland and Albrechtscn, 1995). This calls upon the holistic forces of domain analysis (Garfield, 1992). The present text does not go so far as to explore the perception of human beings regarding visual representation, an interesting matter that is taken up by *Polanco, Francois and Keim* (1998).

From the bibliometric standpoint, the holistic vision in domain analysis is given by the authors of the scientific community. It is the authors themselves, from their respective realms, who constitute and construct part of the discourse of that overall domain. They are responsible for their background, their interests, the relations and interactions among domains. And all this takes place through language, that is through the references of the bibliographic citations in their works. The objective vision in domain analysis is thus given through citation, but by a variant used to generate bibliometric maps or domain representations: cocitation.

We will not deny the fact that author citation can bear a certain degree of subjectivity and intentionality at its core, yet these limitations disappear when cocitation is put to use in a vast realm. Why? Because the different author viewpoints, as seen through cocitation, result in a data set where the opinion of one subjective author cannot prevail. What prevails is the consensus of the bulk of authors. Moreover, cocitation affords invaluable information about how these authors, as experts in a domain, perceive the interconnectivity of their paths of study or interest by means of the studies published to date. In this way, a particular thematic area or research front can be quickly spotted by authors or researchers somewhat familiar with the domain.

This general and objective vision derived from the bibliometric focus, with respect to domain analysis in this case, has been approached by a number of authors dedicated to the analysis and visualization of domains, though until now it had not been postulated in such a patent manner. The finality has thus far been more oriented toward a justification of the maps or visualizations obtained, than to the subsequent analysis of the domain portrayed therein. For instance, Small considers cocitation to represent a relationship established by the citing authors; and so when its strength is measured, what we are doing is calculating the degree of relation or association among documents, just as the community of citing authors sees it. And due to this relevance of author citations, the patterns of relationship can change over time just as the co-occurrences of citations change, making clear how a field evolves in a certain direction or directions.

This can lead us to a more objective manner of looking into the structure of specialized fields of science, as the changes produced in cocitation patterns, over time, give indications of the diverse mechanisms behind the development of a given field (Small, 1973). For *Franklin* and *Johnston*, the bibliometric model based on cocitation, while grouping authors, works, documents or journals according to thematic affinities, allows for a measurement of the interaction among the different research fronts, to the point where we can turn the set into a hierarchy of specialized areas related amongst themselves (Franklin and Johnston, 1988). This hierarchy would not be an a priori classification of knowledge, but rather a self-organizing classification on the basis of the work of the scientific community as a whole. *Ding, Chowdhury* and *Foo* sustain that from this point of view – the Sociology of science and the Philosophy of science – cocitation contributes to making manifest the cumulative advancement of science, showing established interactions and creating new ones (Ding; Chowdhury; and Foo, 1999). For White, cocitations automatically group materials, methodologies, and social affinities just as they are perceived by the citing parties. The sense of a greater or lesser affinity between pairs of citations can be seen by counting cocitations – an overall aggregation for which no single citing author could possibly be held responsible (White, 2003).

Finally, for an adequate approach to the bibliometric treatment of domain analysis, according to *Hjørland*, four key factors must be taken into account:

1. The danger of bias that may be introduced by the data as a consequence of the lack of coverage, documental types, classifications established beforehand, and so on must be carefully accounted for at the time of analyzing and interpreting the information within a domain.
2. The second factor is that every bibliometric map is determined by the patterns of citation of each discipline.
3. The third factor lies in the holistic methods that should be used by researchers when data are analyzed.
4. The fourth and final essential ingredient is the dynamic character of epistemological bases of science.

Yet there is also a fifth necessity that is not mentioned by *Hjørland* and which deserves mention and reflection at this point. The party making the interpretation should be equipped with some previous knowledge of the domain at hand, for example the history of science, the sociology of science, and so on, as this conforms the foundation for an adequate reading of the evolution and the paradigmatic changes occurring within the domain, if any. This fifth factor reinforces the bibliometric perspective, and it serves to close the circle of the holistic vision, as it connects the knowledge provided by the discourse of the pertinent communities with that of the individual, who, whether or not a member of that community, tries to analyze it.

3.2 Social Networks

In spite of their widespread acknowledgment in so many fields, social networks and their analysis are a practically unknown approximation in both theoretical and methodological spheres in the area of Documentation. This trend is changing, however. Social network analysis may stand as a quantitative and qualitative leap in the representation and analysis of the structure of all sorts of scientific domains, whether they be geographic, thematic, or institutional.

In the works of *Liberman* and *Wolf*, the scientific community structures its relations according to models of social networks, where the nodes or actors represent individuals, scientific disciplines, and so on, and the links are the knowledge exchanged by these actors (Liberman and Wolf, 1997). For a detailed review of what social network analysis and graph theory means, we suggest the work of *Wasserman and Faust* in 1998.

3.2.1 Basic Notions

Over time, social networks have developed a terminology of their own, which has grown in parallel with the great variety of studies carried out under different disciplines and perspectives. Hence, beyond some notions that have earned general consensus, there is a proliferation of varied concepts and ideas harbored under apparently well-established terms. This leads to much confusion with regards to key concepts for the analysis of social networks (Herrero, 1999). For this reason, we very succinctly expound the basic terminology of this theory below.

Actor

In social networks, there are two fundamental elements for comprehension. One of these is the actor, a name given to each of the entities or objects of study forming part of the network. In a social network graph, the actor may also be referred to as the node, vertex, or point. The actor does not necessarily have to represent a concrete unit or individual; it may also be a company, institution, or social group.

Link

It is the other key element in the graphic analysis of networks, as the element in charge of connecting one or more actors with others (Fig. 3.1). This may also be called a connection or line, and it may be directional or non-directional, depending on whether it indicates the orientation – from one actor to another – or does not. In the first case, the link is called an arrow or arc (directional link), whereas in the latter case it is non-directional or reciprocal. Links may or may not be weighted, depending on whether they indicate the degree of connection in numerical terms.

There is a special type of link, the self-link or loop, which is produced when an actor makes reference to itself (Fig. 3.2).

Fig. 3.1. Actors and non-directional link

Fig. 3.2. Loop or autolink

Group

A finite set of actors and links, which, for theoretical, conceptual, or empiric reasons, are treated as a closed grouping of individuals (Fig. 3.3).

Subgroup

Subset or finite grouping of actors and links that is part of a greater unit or network (Fig. 3.4).

Relation

Set of links that exist between or among actors of a group or a specific set of actors.

Adjacent Actors

Actors that can be found in direct relation or connection via a link.

Fig. 3.3. Group of actors in a network

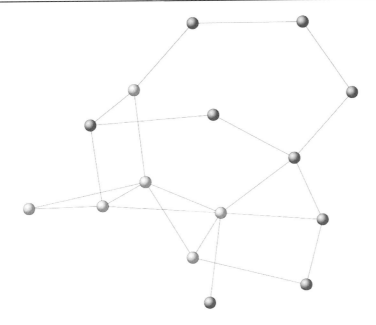

Fig. 3.4. Subgroup of light nodes in a network

Direct Connections

Those produced between adjacent actors, that is, with no intermediary node. They may also be referred to as direct links. For example, the relation of adjacency – discontinuous – existing between actor two and its neighbors in Fig. 3.5.

Neighborhood

Set of actors with which a given actor or node is adjacent (Fig. 3.6).

Indirect Connections

Those made between non-adjacent nodes, through intermediary actors. They can also be called indirect links.

Path

This is the sequence of links and actors that connect two non-adjacent actors, without repeating any of them. The length of the path is determined by the number of links. It makes manifest the existence of an indirect connection (Fig. 3.7).

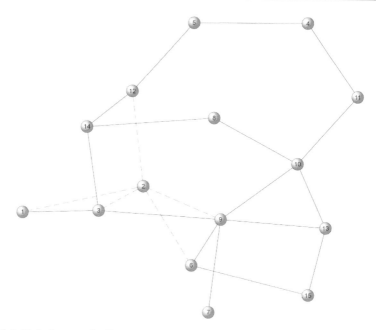

Fig. 3.5. Relations and adjacent connections to actor two

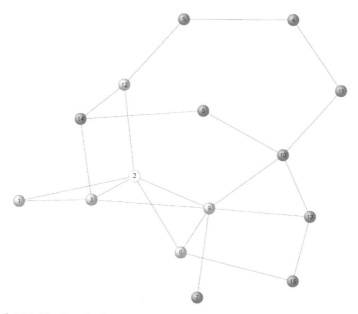

Fig. 3.6. Neighborhood of actor two

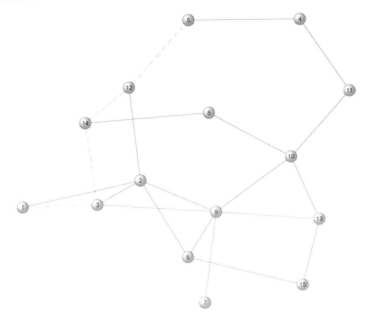

Fig. 3.7. One of the possible paths between actors one and five

Geodesic Distance

It is the shortest path between two nodes or actors of the network, and can also be denominated geodesic length, or simply distance (Fig. 3.8).

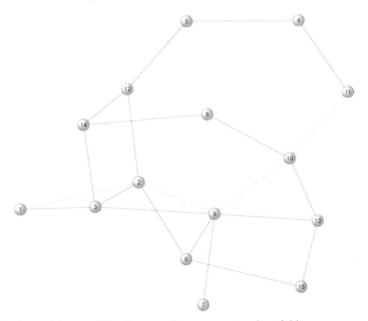

Fig. 3.8. One of the possible distances between actors 1 and 11

Diameter

It is the longest path between two specific nodes or actors (Fig. 3.9).

Isolated Actors

Actors that have no link or relation with any other actor in the network (Fig. 3.10). They may also be called disconnected actors.

Connectivity of a Graph

A graph is said to be connected if there exists a path between each pair of nodes; if not, the graph is said to be disconnected (Fig. 3.11).

Components

This name is given to each one of the subgraphs or subgroups that make up a network.

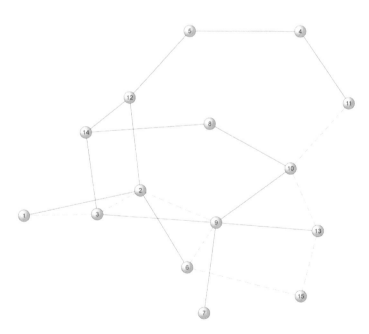

Fig. 3.9. Diameter between actors 1 and 11

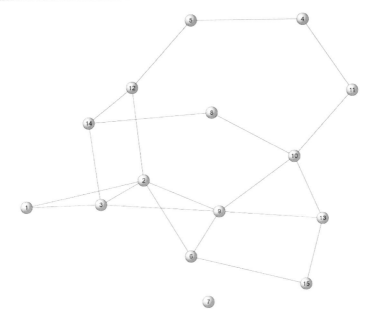

Fig. 3.10. Network with an isolated actor

Fig. 3.11. Connected graph, beside a disconnected graph with two components

Cutoff Point

A node or actor is considered to be the cutoff if, by eliminating that node, and therefore its links as well, the graph is left disconnected, or increases its number of components. Other denominations used for the cutoff points are intermediary, and broker (Fig. 3.12).

Bridge

It is a critical element in the connectivity of a graph. If, by eliminating a specific link between two actors, the graph becomes disconnected, or else increases its number of components, that link is known as a bridge of the network (Fig. 3.13).

Fig. 3.12. Intermediary node or actor in a network

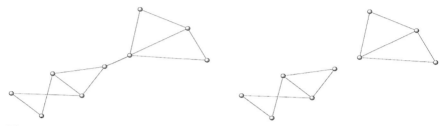

Fig. 3.13. Link that acts as a bridge between two components

Measures of Centrality

These are measurements destined to detect and identify the most important or central actors of a network. This type of measure is based on graph theory. The idea of centrality does not refer to the position of an actor, but rather to its degree of integration or cohesion in the network. Following *Freeman* (1979)**,** the measures of centrality are of three sorts: degree, closeness, and level of intermediation, also known as betweenness. Within these three we can distinguish the following:

a. Centrality of Degree. Also known simply as degree, it is the simplest unit of centrality. It is defined as the number of direct links that an actor has. An actor with a high degree of centrality will have a broad neighborhood, will occupy central positions, will be more visible, and will become an important element for the interconnection of the network.

b. Centrality of Closeness. This type of measure is based on nearness or distance, measuring just how close an actor is to the rest of the actors of the network. The more central the situation of an actor, the greater its capacity of interaction with the rest of the actors.

c. Centrality of Intermediation or Betweenness. The interaction between two non-adjacent actors may depend on other actors in the network, and particularly on those situated on the path of the corresponding non-adjacent nodes. This type of actor or intermediary of paths between actors serves to control the interactions of the network. The centrality by intermediation or betweenness measures the degree to which a given actor – intermediary – forms part of the shortest path or geodesic distance between other actors.

3.2.2 Concept of Social Network

Many of the terms habitually used in social network theory and analysis proceed from the field of Anthropology. Indeed, it is precisely an anthropologist, *Barnes* (1954), who holds the honor of creating the concept of social network. For Barnes, it meant a set of ties that join members of a social system throughout and beyond social categories and closed groups.

According to *Wasserman* and *Faust* (1998), a social network consists of a finite set of elements – actors – and the relations defined among them, where the presence of the latter is a critical and defining characteristic of the network: what is most important in social networks is not the individual, but the structure, defined as the set of individuals and their connections.

Molina, *Muñoz* and *Losego* (2000) affirm that social networks are centered on the identification and analysis of structures, on the basis of the relationships existing between certain elements, regardless of their attributes or characteristics. It is assumed that these structures exert some sort of influence on the behavior of the elements that make up the system.

There is no overall agreement amongst authors dedicated to the study of social networks as to their definition; yet a wide consensus surrounds the most important principles of social networks, which can be summed up as four (Wasserman and Faust, 1998):

- Individuals and their actions are contemplated as independent elements.
- The connections between individuals are studied as chains of transference.
- The networks centered on individuals show a network structure in which these appear as a source of opportunities or limitations for individual action.
- The network models conceptualize structures as if they were fixed patterns of relations among individuals.

3.2.3 Brief Historical Review

The roots of social networks can be found in the 1930s, in Psychiatry and in Social Anthropology (Moreno, 1934), which separately initiated the study of small groups of individuals by means of their components: actors and links. Yet the true growth spurt in their development can be traced to Psychology, which introduced measures designed to obtain the patterns of the social connections that link sets of actors (Freeman, 2000a). The main aim was to detect social groups – actors closely interrelated – or else social positions – actors within a social system that are related in similar ways. In order to represent the connections between the actors and apply measures with which to infer patterns of behaviour, matrixes of coinciding data were used. While rough at the edges and offering only limited information at first glance, this approach gradually took shape as the most adequate method for studying the relations between individuals and groups.

The graphic representation of information for the analysis of this type of pattern was introduced by *Moreno*: "…the sociogram is more than a mere system of representation… it is a method that makes possible the exploration of sociometric events, where the particular emplacement of each actor and its interrelations with other actors can be shown. To date, it is the only possible scheme for carrying out the structural analysis of a community."

The sociogram affords the advantage of transforming mathematical information contained in numerical matrixes into visual information, or graphs. From the viewpoint of abstraction, visual information promises huge benefits with respect to purely numerical information, as it enhances transmission of the structural basis of the network and highlights the relevance of the different actors.

In a matrix, the social actors are represented as lines (cases) and columns (variables); while the links are the values existing in the correspondence between these two (Table 3.1).

In a sociogram, the actors are depicted as nodes or vertices, and the links are the relations existing between actors who interact in mutual fashion (Fig. 3.14).

The sociogram developed by *Moreno* marked the onset of sociometrics, which in turn gave rise to the analysis of social networks through the measurement of interpersonal relationships with reference to reduced groups.

Table 3.1. Matrix of data cocitation

	1	2	3	4	5	6
1	0	1	0	0	0	0
2	1	0	0.525	0	0.33	0
3	0	0.525	0	0	0.217	0.609
4	0	0	0	0	0.139	0
5	0	0.33	0.217	0.139	0	0
6	0	0	0.609	0	0	0

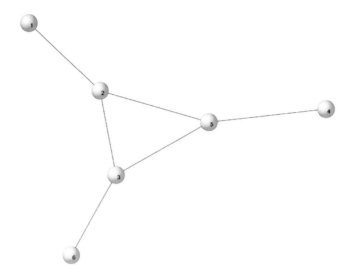

Fig. 3.14. Social network of nodes and links, or sociogram

In the early years, sociograms were drawn by hand, so that results were largely determined by the artistic dexterity of the person in charge of the network representation. With the development of computers since the 1960s, however, researchers increasingly liberated from the processing of huge volumes of data became able to produce results comparable with those obtained manually, but with greater scientific validity and reliability (Corman, 1990).

The incorporation of computer technology to social networks meant a real boost, allowing, in the words of *Freeman* (2000b),

- the representation and analysis of complex structures;
- the construction of graphs in two or three dimensions; and
- the application of techniques of MA, such as Factor Analysis (FA) to social networks, for the positioning and spatial reduction of representations.

From the 1970s onward, with the progress of discrete mathematics and, in particular, graph theory, a conceptual framework arose, lending the analysis of social networks a "formal language for the description of networks and their characteristics" and "the possibility of interpreting the data of a matrix as formal concepts and theorems, which can be directly related with the characteristics unique to social networks" (Scott, 1992).

Just as the sociogram constituted the means of graphically representing a matrix of relational data, the language of graph theory took hold as another much more general form of doing much the same. Graph theory is not simply another mathematical approach, but stands rather as the turning point of most ideas behind the analysis of social networks. Thus we can say that a sociogram is the graphic representation of a data matrix that receives the name of graph or graphics, which by virtue of graph theory is transformed into concepts and theorems, giving as a result a social network made up of actors and links.

After the 1980s, with the introduction of personal computers, the graphic representation of social networks was even closer at hand. Computer screens helped researchers apply the different techniques of grouping and detection of social positioning on a medium other than paper. At the same time, these sociograms could be more easily validated by independent researchers, to contrast the results of studies.

From then till now, social networks and their analysis have been used in a wide array of research areas, such as work mobility, the impact of residence on the individual, political and economic systems, decision-making, social support, communities, problem-solving groups, diffusion, corporative relations, beliefs, perception and cognition, the market, the sociology of science, power and influence, consensus and social influence, formation, communications between computers, organizational structures, health and disease (AIDS), article or journal citation and cocitation studies, the visualization of scientific domains as interfaces for information retrieval, and – as of very recently – in the analysis and structure of science.

3.2.4 Social Network Analysis

The appearance of social networks as a tool for studying the social behaviour of small groups of individuals, the development of the sociogram as a technique for the graphic representation of such relations, and the appearance of graph theory as a conceptual framework for the description and analysis of social networks shared a common objective: to unbury, in a deep but appropriate manner, the underlying structures of social networks. But where was this analysis to lead us?

Cassi (2003) believes that, because the social network is based on the interdependence of actors and their actions, its analysis should be focused more heavily on the relations than on the characteristics or attributes of the elements it comprises.

For *Freeman* (2000a), the analysis of social networks is an interdisciplinary specialty of the behavioral sciences, whose objective is to observe the interdependent elements under study (the actors) and how the interactions or relations taking place among them affect each actor. The analysis of social networks implies the elaboration of theoretical and empiric models in order to discover patterns of relations among actors and the precedents and consequences of such patterns.

Molina (2001) affirms that the analysis of social networks studies specific relations between a defined series of elements; and that unlike traditional analyses, it focuses on the analysis of the relations and not on the attributes of the elements. It deals, therefore, with relational data and is capable, by this means, of identifying and describing a structure in an operative way, not metaphorically.

Rodriguez (1995), on the other hand, sustains that the analysis of networks attempts to explain the behavior of network elements and of the system as a whole. This implies rejecting attempts to explain social processes and individual conducts based exclusively on the attributes of the actors. In the face of this type of individualistic analysis, so typical of the social sciences, network analysis tries to explain the behavior of individuals as the result of their social relations.

Wasserman and *Faust* (1998) are of the opinion that network analysis integrates theories, models, and applications that are expressed in terms of conceptual relations. That is, they consider that relations are the fundamental component of the network theory.

From these definitions, it is derived that network analysis provides a new method for the examination of processes with respect to those used up until now. The main difference resides in that it is not based on an individualistic analysis of the characteristics of the authors, but rather is elaborated with relational information concerning the actors who constitute the network structure.

The analysis of social networks, also known as structural analysis, has left many a researcher with a wrinkled brow. As *Welman* (1988) puts it, some underestimate it with the argument that it is mere methodology and lacks sufficient merit for dealing with substantive problems; others shy from its strange terms and techniques, not having played with blocks and graphs since primary school; others cut out a part of the whole, asserting, for instance, that their studies about the structure of classes do not require concentrating on the ties of friendship emphasized by network analysis; others disdain it as something that is not at all novel, implying that social structure has always been an object of study; others hook onto variables such as network density, as if to compress the variance explained; others, attracted by the possibility of studying non-hierarchical structures in a non-group context, broaden structural analysis into an ideology of networks to fight for egalitarian and open communities. Some even use network as a verb, to advocate the creation and deliberate use of social networks with desirable aims, such as employment or community integration.

Such pseudo-conceptions have risen due to the fact that many analysts and professionals have misused social network analysis as a catchall of terms and techniques. Still others have frozen the term to reduce it to a method, while others smooth it into a metaphor. Many have limited the power of this focus by treating all the units as if they had the same resources, all the ties as if they were symmetrical, and all the contents of ties as equivalents. Nonetheless, structural analysis does not derive its power of partial application from this or that concept or measure. It is a comprehensive and paradigmatic form of considering the social structure in a serious manner, from the direct study of the way in which the patterns of union assign resources in a social system. Therefore, its force stems from the integrated application of theoretical concepts, manners of obtaining and analyzing data, and a growing cumulative corpus of substantive findings.

3.3 Scientography

Scientific information is found scattered over disciplines which, for the non-expert, or even sometimes for the experts themselves, bear little or nothing in common. When one studies a domain pertaining to a specific field of knowledge along traditional lines, he may be left with the sensation of not grasping the domain in its entirety.

Science maps can be very useful for navigating around in scientific literature and for the representation of its spatial relations (Garfield, 1986). They adequately depict the spatial distribution of areas of research, while at the same time offering additional information through the possibility of contemplating these relationships (Small & Garfield, 1985). From a general viewpoint, science maps reflect relationships between and among disciplines; and the positioning of their tags clues us into semantic connections while also serving as an index to comprehend why certain nodes or fields are connected with others. Moreover, these large-scale maps of science show which special fields are most productively involved in research, providing a glimpse of changes in the panorama, and which particular individuals, publications, institutions, regions, or countries are the most prominent ones (Garfield, 1994).

The construction of maps from bibliometric information is also known as scientography. According to Garfield, this term was coined by the person in charge of basic research at the Institute of Scientific Information (ISI), George Vladutz, to denominate the graphs or maps obtained as a consequence of combining scientometrics with geography (Garfield, 1986). Although "scientography" not a widely familiar term, possibly due to the proliferation of terms such as "domain visualization" or "information/knowledge visualization" that make reference to similar notions, in our opinion it is the most adequate term for describing the action and effect of drawing charts of scientific output.

And so scientography, by means of its product known as scientograms, has become a tool and method for the analysis of domains in the sense used by *Hjørland & Albrechtsen* (1995), consolidating the holistic and realistic focuses of this type of analysis. It is a tool in that it allows the generation of maps; and a method in that it facilitates the analysis of domains, by showing the structure and relations of the inherent elements represented. In a nutshell, scientography is a holistic tool for expressing the discourse of the scientific community it aspires to represent, reflecting with accuracy the intellectual consensus of researchers making up that community, on the basis of their own citations of scientific literature.

3.4 Scientography and Domain Analysis

 Scientography can be considered a tool and method for the analysis of domains, supporting the holistic approach and the objective of this type of analysis. It is a tool in that it permits the construction of bibliometric maps, and it is a method because it facilitates domain analysis by showing in graphic form the structure and relationships of the elements represented in the domain. These maps reflect the bulk of opinions and viewpoints of the persons who make up the given community.

Scientography is also a holistic method for another reason: because it allows the domain to be analyzed in light of the community discourse; and it is an objective method because through it we are able to analyze the non-subjective structure, formed by the intellectual consensus of the component relations among its elements.

This synthesis between the tool-methodology and the holistic-objectivism of scientography is further strengthened by the use of social networks in the graphic representation of bibliometric maps. These make it possible to discover and demonstrate theories about the graphs themselves and, consequently, about the models they represent. This makes them highly useful for the formal representation of social relations, as well as for detecting and quantifying their structural properties. Researchers can, through them, discover otherwise hidden patterns and trends.

In the area of Documentation, the visualization of information by means of the graphic representation of social networks has become one of the main techniques with which to make manifest the intellectual relations and the structure of scientific knowledge. This new approach not only magnifies the possibilities of analysis of traditional domains, or of scientific disciplines; it also constitutes a key tool with which to study the interaction and evolution of science by means of the disciplines and specialities it embraces.

In summary, we can say that scientography – understood as the bibliometric mapping of social networks – is a tool and a method, which can be applied to reveal the explicit acknowledgment of some authors by others, and therefore the overall interactivity and evolution of any particular domain of study.

4 Methodological Aspects Previous to Scientography

White and *McCain* (1997), in their work *Visualization of Literature*, distinguished five models of information: bibliographic, editorial, bibliometric, for users, and synthetic.

Nowadays, as a consequence of the development of information technology and the techniques for the processing and representation of information, the borderlines between these models are somewhat fuzzy. The model used by researchers today could be described as a "user meta-model". It is a user method in that it represents a reduction of the informational base, responding to queries, needs, and user profiles, by means of complex computer processes. And it is a meta-model in the sense that it contains metadata such as author, title, date, and category that can be utilized to show the relationships between the units of study as well as for the construction of maps. Furthermore, these data contain and can provide bibliometric information: number of citations, cocitation, impact factor, and so on. This user meta-model is closely associated with the processes and techniques used for the generation of maps or visualization of domains (Börner; Chen; and Boyack, 2003).

We offer a synopsis of these processes with their possible variations in Fig. 4.1. This is a simplification based on the proposal made by the authors cited just above in which we have eliminated some elements and placed special emphasis on others, precisely those that will be used in our attempt to visualize a vast scientific domain (to be described in detail in Chap. 6).

The stages shown in the figure are now presented one by one.

4.1 Gathering Information

For the structural representation of a scientific domain, it is crucial to first have the information necessary to do so, and second, that this information be adequate for reflecting in a reliable manner the structure of the domain to be represented. Much has been written about the choice of sources, and about the different strategies for locating and extracting information. We

Gather information	Unit of analysis	Unit of measures	Dimensionality reduction and layout of the information	Display

Searches
ISI
INSPEC
Eng. Index
Medline
Research
  Index
Patents
Etc.
Broadening
Por citas
Por términos
	Common
	choices
	Countries
	Topic
	areas
	
	Categories
	Journals
	
	Documents
	Authors
	Terms
	Words
		Valores/
		frequencies
		Attributes (e.g.
		terms)
		Citations
		Co-citations
		By year
		Thresholds
		By counts
			Dimensionality
			reduction
			  Cluster Análisis
			  (CA)
			  Factor Análisis
			  (FA)
			  Componentes
			  Principales (PCA)
			  Multi-dimensional
			  Scaling (MDS)
			  Pathfinder networks
			  (PFNET)
			  Auto-Organizative-
			  Maps, including
			  SOM, ET-maps,
			  etc.
			
			Blockmodeling
			Layout
			  Kamada-Kawai
			  Fruchterman-
			  Reingold
				Interaction
				Browsing
				Pan
				Zoom
				Filter
				Query
				Analysis

Fig. 4.1. Stages in the process of information visualization

will not go back so far as to dig into these considerations, but will depart from the acknowledgment of their importance, as the quality of the scientogram of a scientific domain is directly related to the quality of the information used to build it.

4.2 Unit of Analysis

The most commonly used units of analysis for representing scientific domains are journals, documents, authors, terms, and words or keywords; though recently some newcomers have joined this list: countries, thematic areas of different levels, institutions, and ISI categories.

The choice of one unit over another will depend on the degree or depth of analysis that is desirable in the domain representation; accordingly, each representation will have different facets, adaptable to different sorts of analyses and access.

It is not commonplace to find domain visualizations in which several units of analysis converge. Yet it is feasible, and the rules of the game do not prohibit attempts in this direction. Possibly, however, the tremendous cognitive effort required to analyze representations of domains with these characteristics dissuades some researchers. It is, in fact, possible to use units with a greater capacity of agglomeration to represent the general structure of a domain, and smaller units with less agglutinating potential to go into lower levels of specificity.

One aspect related with the units of analysis that we must take into account is the amount of information they have available, and hence the total size of the domain to be represented. If the number of variables or items with which we are going to work is very limited, we will be able to build displays of domains with very small units, such as words or descriptors. If this is not the case, we will need to consider using larger units of analysis, such as documents or authors. But if the amount of information to be accounted for is great or very great, we will need to resort to units of analysis capable of containing smaller units, as is the case of journals that group documents, authors, and terms, or of categories, which embrace all at the same time. This consideration is nothing new in the field of information visualization, and it stems from a physical limitation inherent to the representation of information in a reduced space of low resolution. For this reason, authors such as *Tufte* (1994) have been studying this matter for over a decade, to come up with a variety of solutions.

4.3 Units of Measure

The purpose of the units of measurement is to quantify the relationship between each one of the members of the unit of analysis selected with the rest of its components. The result is a series of multidimensional data matrixes that make manifest the existence of these relationships and the extent of them.

As *Börner, Chen* and *Boyack* (2003) acknowledge, the units of measure have been studied at length and in depth by *White* and *McCain* (1997), and like them, we therefore find it most appropriate to use the terminology both propose:

> We use certain technical terms such as intercitation, interdocument, coassignment, co-classification, co-citation, and co-word. The prefix "inter-" implies relationships between documents [or units]. The prefix "co-" implies joint occurrences within a single document [or unit]. Thus, intercitation data for journals are counts of the times that any journal cites any other journal, as well as itself, in a matrix. (The citations appear in articles,

of course.) The converse is the number of times any journal is cited by any other journal. The same sort of matrix can be formed with authors replacing journals. Interdocument similarity can be measured by counting indicators of content that two different documents have in common, such as descriptors or references to other writings (the latter is known as bibliographic coupling strength). Co-assignment means the assignment of two indexing terms to the same document by an indexer (the terms themselves might be called co-terms, co-descriptors, or co-classifications). Co-citation occurs when any two works appear in the references of a third work. The authors of the two co-cited works are co-cited authors. If the co-cited works appeared in two different journals, the latter are co-cited journals. Co-words are words that appear together in some piece of natural language, such as a title or abstract. Bother "inter-" and "co-" relationships are explicit and potentially countable by computer. Thus, both might yield raw data for visualization of literatures.

From the standpoint of domain visualization, there are various units of measure that can reveal the relationships existing among units of analysis and thus evidence the intellectual structure they constitute. This is the case of citation, bibliographic coupling, and cocitation. Of these, cocitation stands today as the most widely accepted technique, while also the most often used for the visualization of scientific domains.

Cocitation can be defined as the frequency with which two units of measure – authors, documents, journals, and so on – are cited by other documents that have been published at a later date. As Small puts it, a more formal definition of cocitation would be as follows: if A is the set of documents that cite document a, and B is the set that cite document b, then $A \cap B$ is the set that cite both a and b. The number of $A \cap B$ elements, that is, $n(A \cap B)$, is the frequency of cocitation. Hence, the relative frequency of cocitation would be defined as $\dfrac{n(A \cap B)}{n(A \cup B)}$ (Small, 1973).

4.3.1 Document Cocitation

Also known as work co-referencing, it is the precursor of the cocitation of all other units of measure. It is defined as the frequency with which two documents or works – books, journals, articles, and so on – are cited by other works or documents published thereafter.

The basic notion behind document cocitation revolves around the principle that two documents are cocited by a third because they have some sort of mutual relationship, and this relation will be more significant when the number of cocitations is high. The relations among documents can be used to represent the scientific structure of a domain or even to detect its research fronts. Despite providing a granular analysis of the sciences, finer than that obtained with other units of measure, it is not very widely used for domain representation. However, the first maps of science, and some later ones, have been created using these units (among others, Small and Griffith, 1974; Griffith et al., 1974; Garfield, 1981; Small and Garfield, 1985; Small, 1997; Garfield, 1998).

4.3.2 Author Cocitation

Out of all the methods for coassignment or co-occurrence, that of authors is the most widely used. Author cocitation takes place when one author cites, in a new document, any work by another author, together with the work of a third, fourth, or fifth author.

This is based on the understanding that works cited in conjunction (cocited) reveal the existence of an intellectual relationship between the cocited authors. The greater the number of cocitations, the stronger that relationship.

Since its appearance (White and Griffith, 1981b; White and Griffith, 1981a; White and Griffith, 1982), Author Cocitation Analysis (ACA) has been characterized by frequent use in the building of two-dimensional representations. In these displays, cocitations automatically group materials, methodologies, and social affinities, as they are perceived by the citing parties. The matrixes that gather the recounts of cocitation over time, with values that vary from zero to perhaps thousands, can be the basis of a domain visualization (White, 2003). ACA has been carefully explained and graphically synthesized by *McCain* (1990).

The past 20 years have witnessed the application of several techniques for the construction of ACA-based visualizations. In data entry, cocitation values in a pure state have been used, as well as the recount of the number of pairs of authors standardized through some type of similarity measure such as the Pearson correlation coefficient, or that of Salton, or the cosine. For the spatial distribution of the information displayed, techniques have been sought in MultiDimensional Scaling (MDS), clustering, FA, Self-Organizing Maps (SOM), geographic maps, and PFNET. Of these, the two-dimensional representations obtained with PFNET and effected with pure cocitation values, then visualized through spring-embedder-type programs, are the ones that appear to offer the best results (Lin; White; and Buzydlowski, 2003) as we will see later on.

While by far the most often adopted technique, this one has its drawbacks as well. The weakest point is the need for a subjective interpretation of results, as it calls for the person interpreting to have a considerable knowledge of the domain of study. At the same time, complaints also surround its difficulty in identifying groups within the domain maps (Ding; Chowdhury; and Foo, 1999).

It is true that ACA is a tool with great potential for the display of the intellectual structure of the different disciplines within science, as it shows and validates the intellectual structure of the domain it represents, by means of the consensus of the main authors involved therein. As proof we have the numerous works focusing on it as the unit of analysis (White and McCain, 1998a; Chen and Carr, 1999b; Chen; Paul; and O'keefe, 2001; White; Buzydlowski; and Lin, 2000; White, 2000; White, 2001; Lin; White; and Buzydlowski, 2003; White, 2003), and a long etcetera.

4.3.3 Journal Cocitation

Here the basic idea is very much the same: the more the citations of two journals, hand in hand, the greater their mutual relationship must be.

As with documents and authors, journals can also be distinguished by diverse characteristics: a wide or limited specialization of the subject matter, methodological orientation, institutional affiliation, professional prestige, and other attributes. Within journal cocitation, that of articles connects the journals where specific articles have appeared. So, two journals will be cocited when at least two of their published articles, which may belong to one or different journals, appear referenced in the bibliography of a new article. Accordingly, the journals with a general character will have greater possibilities of cocitation with a large number of publications, whereas the more specialized ones will be cocited with other more focused journals (McCain, 1991a).

Data on cocitation proceeding from journal articles gives us crucial information about researchers and the disciplines where they publish, that is about the thematic coverage of the journals in which authors publish. Journal Cocitation Analysis (JCA), like ACA, holds great interest because of the possibilities it lends for building displays that bring to surface the underlying scientific structure of a given domain, which, in the case of journals, makes reference to the discipline(s)

The JCA has given rise to studies of the structure of different scientific domains: Psychology (Doreian, 1985), Geography (Doreian, 1988), Communications (Rice; Borgman; and Reeves, 1988), Economics (McCain, 1991a); Genetics (McCain, 1991b); Medical Informatics (Morris, 1998), Computer Engineering (Marion and McCain, 2001); Cardiovascular Medicine (Jarneving, 2001), and Library and Information Science (Bonnevie, 2003), among others.

Likewise, it can be applied to a discipline on the whole, or else to the bibliographic references of a single journal (Moya-Anegón and Herrero-Solana,

2001), thereby representing the intellectual profile of that journal itself, as reflected in the articles it has put into print.

4.3.4 Cocitation of Classes and Categories

Class and category cocitation was first proposed by a group of researchers from the University of Granada (Grupo SCImago, 2002) as a useful technique for the construction of maps of vast scientific domains. It is based on the extension of the traditional cocitation scheme and on the use of units of analysis greater than those used previously (Moya-Anegón et al., 2004).

The superior units proposed are, in ascending order: the Journal Citation Report (JCR) categories (The Thomson Corporation, 2005a) – in its versions Science Citation Index (SCI) and Social Science Citation Index (SSCI) – and the classes of Spain's *Agencia Nacional de Evaluacion y Prospective* (ANEP, 2005). The cocitation scheme proceeds in this fashion.

The research community widely accepts, in the area of information visualization and display, that the frequency with which any two documents are cited in conjunction, or cocited, represents the degree of affinity of the two according to the perspective of the citing author(s) (Moya-Anegón et al., 2004). The cocitation value reveals the number of times that two cocitation units, be they authors, journals, words, JCR categories or ANEP classes, have been cited together in posterior works.

The JCR, depending on the disciplines that each journal claims to embrace, assigns each to one or more thematic categories. The Journal of the American Society for Information Science and Technology (JASIST), for instance, is ascribed by the JCR to the subject areas Computer Science and Information Systems and Information Science and Library Science. Therefore, if we follow the order of the model of cocitation of Fig. 4.2, with

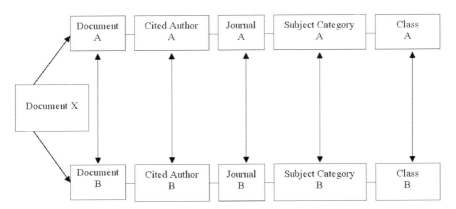

Fig. 4.2. Scheme of the workings of class and category cocitation

authors, documents, journals, JCR categories, and ANEP classes, we deduce that the two categories can be conveniently used as units of cocitation and analysis to represent the intellectual reality of a domain. Similarly, if documents A and B of Fig. 4.2, which appear in the journal (JASIST), are cocited by document X, that cocitation makes manifest a relationship between the categories Computer Science and Information Systems and Information Science and Library Science, to the extent or intensity marked by the number of times the two categories are cocited. Then, if document A was published in that journal, and has been assigned the corresponding categories mentioned above, and document B has been published in the European Journal of Operational Research, assigned to JCR categories Management and Operations Research & Management Science, we become aware of the relationship between categories Computer Science and Information Systems and Information Science and Library Science, and Management and Operations Research & Management Science. The intensity of this relationship, again, will depend on the number of times that these categories are cited in conjunction, whether by name or else because of cocitations of the articles they have published.

In following the reasoning used for the JCR categorizations, the ANEP classes are also well suited to the task of representation of the intellectual structure of a domain. And so, to sum up, the cocitation model explained here assumes that the references to document x evidence a cocitation relation between the cited documents and their respective authors. By analogy, we can say the same of the journals on whose pages these documents appear, and of the subject categories that the JCR gives to each. This is a clear and easy way to extrapolate document cocitation to broader panoramas or even to higher levels of groupings.

To demonstrate the intellectual relations existing among the different disciplines acknowledged in the SCI, SSCI and Arts & Humanities Index (A&HCI) (or, in other words, between the pure sciences, the social sciences, and arts and humanities, as if they all were a single entity) and to resolve the problem of standardizing the different degrees of citation among disciplines that appear in the various databases, *Moya* and his research team follow in the footsteps of *Small* and *Garfield*, and their work, *The Geography of Science: disciplinary and national mappings* (Small and Garfield, 1985). They standardize class and category cocitation, dividing the cocitation value by the square root of the product of the frequency of the citations of the cocited categories. The standardized cocitation measure is given in Eq. 4.1 (Salton and Bergmark, 1979):

$$MCN(ij) = \frac{Cc(ij)}{\sqrt{c(i) \cdot c(j)}} \qquad (4.1)$$

where:

Cc is Cocitation
c is citation
i, j are categories.

4.4 Reduction of the *n*-Dimensional Space

The multidimensional matrixes to which we made reference under Sect. 4.3, which were in charge of showing the relations among the different elements that constitute a domain, need to be interpreted. This is necessary because, due to their informative wealth, the human mind requires much effort and energy to decipher such matrixes, when this is in fact possible. In most cases it will be necessary to resort to methods that allow for the transformation of one *n*-dimensional space into another having two or three dimensions.

This section will profile some of the techniques applied for the reduction of *n*-dimensional space. Specifically, those that can prove useful in the representation of the structure of a domain on paper or on the computer screen. We shall distinguish three major groups: those of a multivariate statistical nature, those of connectionist origins, and those based on social network analysis.

4.4.1 Multivariate Analysis Methods

As we have been seeing, the complexity of scientific phenomena forces researchers to gather multiple measures in order to properly capture their natures. This brought about the rapid implantation of multivariate or multi-variable methods, which allow the simultaneous analysis of broad sets of variables. The part of statistics or data analysis that deals with methods for analyzing multiple variables is known as multivariate analysis, or MA. In a broad sense, MA can be described as the set of methods that analyze the relationships between a reasonably large number of measures, taken for each object or unit of analysis, in one or more samples at the same time (Martínez Arias, 1999). There are many types of MA, and we will look at just three.

Cluster Analysis

The term "cluster analysis" was coined in 1939 by *Tyron* (1939). Since, it has also been called analysis of conglomerates or numerical taxonomy. It consists of a multivariate statistical technique whose end purpose is to divide a set of objects into groups, or clusters, in such a way that the profiles of the objects in one same group will bear a great resemblance to each other (internal cohesion of the group) and the profiles of the objects of different clusters will be different (external isolation of the group).

By means of cluster analysis, or clustering, we can reduce the volume of information by grouping data of similar characteristics. Clustering leave us with a two-dimensional image called a dendrogram, which shows the different groupings of objects on the basis of the underlying relations, contained in the data matrix. There is an array of some 150 techniques for hierarchical clustering, depending on the principle of agglomeration applied. Yet all share two essential elements: the function of distance, and the rules of agglomeration (Faba-Pérez; Guerrero Bote; and Moya-Anegón, 2004). The most commonly used in the Library & Information Science field is the *Ward method*.

Cluster analysis often appears combined with FA and MDS in the context of domain visualization, as we will see shortly. Clustering produces groupings of objects or variables, whereas MDS generates *n*-dimensional visualizations where the variables are displayed and ordered, to show their structural characteristics and internal relations.

Multidimensional Scaling

The MDS consists of a set of techniques sometimes called perceptive mapping, which allows the researcher to determine the objective and relative images that subjects have of a set of objects, and the dimensions on which such judgements are based (Martínez Arias, 1999). It proves very useful when one wishes to set off latent dimensions not visible to the human eye. And also, to compare variables when no clear criteria for comparison exist – although the latter can be a drawback as well, because there are no clear objective criteria for interpreting an MDS image.

The MDS is used to identify the dimensions that best reveal the similarities and distances among variables, and its objective is to generate a map of objects. The applications involve the input of a matrix of similarities or distances, to calculate in an iterative form a series of coordinates in a space of two or three dimensions. In this way, the distances or similarities obtained resemble as strongly as possible those contained in the matrix. In order to ensure the best fit among some distances and others, a statistical measure known as stress is used; it measures the goodness-of-fit between the observed similarities and the calculated ones (Herrero Solana, 1999).

In the field of information visualization, MDS has been widely used for the graphic representation of ACA (Fig. 4.3) (White and McCain, 1998a), domain analysis (White; Lin; and McCain, 1998), the visualization of science maps (Small, 1999), the visualization of research fronts (Fig. 4.4) (Moya-Anegón;

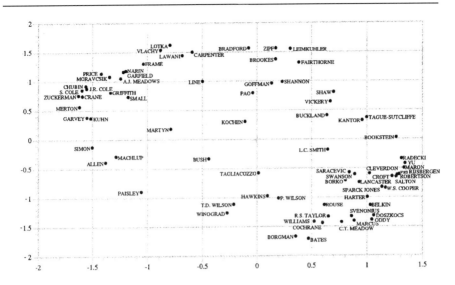

Fig. 4.3. MDS of the 75 main authors in LIS (White and McCain, 1998a)

Fig. 4.4. MDS of the research fronts in LIS (Persson, 1994)

Jiménez Contreras; and Moneda Carrochano, 1998; Ingwersen and Larsen, 2001), the indexing of documents (Ingwersen, 2001), and the evaluation of library catalogs, or Online Public Access Catalogues (OPACs) (Herrero Solana and Moya-Anegón, 2001).

Factor Analysis

The term FA was first used in 1931, when *Thurstone* (1931) developed a multivariate exploratory technique for examining a wide variety of data sets. Its main practical application is to reduce the number of variables, detect the structure by means of its relations, and arrive at its classification. Thus, FA is used as a technique of data reduction and for structural detection (Börner; Chen; and Boyack, 2003).

In essence, we can state that FA attempts to explain the relations existing among the original variables by producing a reduced number of these variables or factors.

Principal Factor Analysis (PFA) is a multivariate technique that used a linear model to explain an extensive set of observable variables in the guise of a reduced number of hypothetical variables called common factors. It sets off the principle or strategy of scientific parsimony or descriptive economy. This principle assumes that variables can be reduced to common factors; which in turn relies on the assumption that each variable is related to a greater or lesser extent with each one of the factors that it contains; which is to say that each factor is present to some varying degree in the variables (Herrero Solana, 1999).

Meanwhile, Principal Component Analysis (PCA), presumes that each factor should explain the maximum amount of initial variability without differentiating between common and specific factors. The basic premise here is that the best means of representing the linear relation between two variables is through a regression curve. The two variables, then, are combined to give way to a third one, designated as the factor. In short, PCA transforms a set of correlated variables into a set of non-correlated variables or components. The new components that arise from PCA represent linear combinations of the original variables and are derived in decreasing order with respect to their importance, according to the degree of variation with respect to the original data. The PCA can be used to reduce the number of dimensions between pairs of variables in order to simplify the graphic representation of the elements included in the matrix. Each one of the new dimensions is called a factor or a principal component, and they go from the first factor or principal component to the last. The first is the one that accumulates the greatest amount of variance, the second a little less, and so on until arriving at the last one (Faba-Pérez; Guerrero Bote; and Moya-Anegón, 2004). Each factor is made up of a certain number of units with a weight, known as factor loading. For example, in Category Cocitation Analysis (CCA), each of the categories that conforms a factor is granted a specific load. In addition, each factor can be characterized or tagged, bearing in mind the categories that reach the established threshold, and the residual factors whose load is under that cutoff are usually not tagged.

The advantages of FA include the fact that it does not force the variables to belong to just one group, as clustering does, but rather allows them to be classified in different factors, thereby reinforcing the idea that what really matters for a given work is its universal character.

In FA, the total number of factors that can possibly be extracted is equal to the total number of variables with which one works. Nonetheless, and due to the need to reduce the number of dimensions in order to interpret the information, only those that surpass a determined threshold value will be taken into account. This cutoff value is a number that expresses the percentage of variance accumulated by each factor – the *eigenvalue* – with respect to the rest. Generally speaking, the first factors, those with a greater eigenvalue, accumulate a very high percentage of variance, which means that they themselves can represent the totality of the *n*-dimensional space. The rest of the residual factors accumulate little or very little variance, and so they are not usually taken into account, even at the risk of losing information, as the amount they represent is negligible with respect to the first.

Although FA has been mainly used as a technique for the reduction of space and for analysis (White and McCain, 1997; Chen, 1998a; Ding; Chowdhury; and Foo, 1999; Chen and Carr, 1999c; Chen and Carr, 1999b; Chen and Carr, 1999a; Chen, 1999b; Chen; Paul; and O'keefe, 2001; Chen and Paul, 2001), in the context of domain visualization, it has always been used for the detection and identification of groups or factors, contained in visualizations obtained using other methods of spatial reduction. Such is the case of Figs. 4.5 and 4.6, which show how the overlapping use of AF over an MDS graph and a PFNET, respectively, allows one to better detect intellectual groupings and enhance domain analysis by making these displays more informative and more comprehensible.

Figure 4.5 is the representation obtained with MDS of the ACA done for Library Science & Information Science (LIS) in Spain, from 1985 to 1994. The authors are distributed in space according to their affinity, or similarities. The research fronts detected through ACA and FA coincide in the two cases, and they are shown by circles grouping the authors. In light of the groupings and with some knowledge of the activity of the authors themselves, the authors define four research fronts: Libraries, Infometrics, University 1 and University 2.

In the example of Fig. 4.6, in contrast, the different intellectual groups detected by FA stand out, identified by a common gray tone.

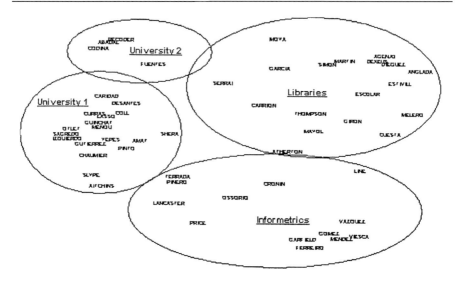

Fig. 4.5. Map of LIS in Spain (Moya-Anegón; Jiménez Contreras; and Moneda Carrochano, 1998)

Fig. 4.6. Groups detected in a PFNET network (Chen and Carr, 1999c)

Blockmodeling

Blockmodels are graphic representations of groupings of redundant or similar actors in a network, clarifying the patterns of relationships among these actors (Fig. 4.7) (Borgatti and Everett, 1992). Blockmodeling aims to favor the recognition of a network's structure and optimize its processing. The re-ordered matrix of blocks is shown in Fig. 4.8.

Blockmodeling starts with an image matrix, which as its name indicates is a matrix that represents the simplified reproduction of the original matrix by grouping the actors into positions and their relations into roles (Fig. 4.9). In the image matrix, each position occupies a cell of the matrix and the value of each will be zero or one, depending on whether or not there are relations between those two positions.

This process is used to get at the essence of social networks by replacing complex information about actors and their relations with simplified equivalent positions and relations. The result can be represented in the form of a matrix using the so-called image matrixes, or else graphically by its blockmodeling. The simplification is achieved through the adoption of a measure of equivalence – normally structural – and the execution of statistical techniques: hierarchical clustering or CONvergence CORrelation (CONCOR), to identify structurally equivalent groups of actors.

	1	2	3	4	5	6	7	8
1	0	0	0	10	0	0	0	0
2	0	0	4	0	11	0	0	0
3	0	4	0	12	5	5	0	0
4	10	0	12	0	0	21	0	0
5	0	11	5	0	0	16	19	15
6	0	0	5	21	16	0	25	14
7	0	0	0	0	19	25	0	23
8	0	0	0	0	15	14	23	0

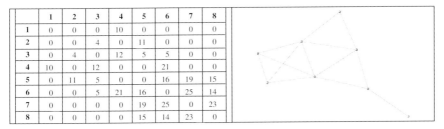

Fig. 4.7. Original matrix and its corresponding network

	1	2	3	4	5	6	7	8
1				1				
2			1		1			
3		1		1	1	1		
4	1		1		1			
5		1	1			1	1	1
6			1	1	1		1	1
7					1	1		1
8					1	1	1	

Fig. 4.8. Re-ordered matrix of blocks

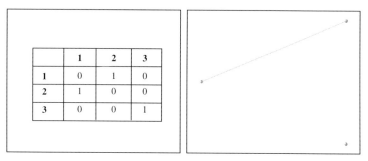

	1	2	3
1	0	1	0
2	1	0	0
3	0	0	1

Fig. 4.9. Image matrix and its corresponding network of blocks

While seldom used in Sociology and Psychology as a technique for the simplification of information, blockmodeling in the field of LIS is altogether unknown, not to mention in the area of domain visualization. The most likely reason is the sparseness of its representations and the lack of information. Besides, it is not a prime technique for detecting the main structure of a domain, as "more than a few authors have used these techniques for detecting already known structural positions, and they have not managed to detect, or identify, such positions" (Wasserman and Faust, 1998).

4.4.2 Connectionist Techniques

Another group of techniques that allows us to reduce dimensions are the connectionist techniques. These constitute a very different approach from the above, as they proceed from the field of Computer Science, specifically from a line of research known as soft computing. This line takes on problem-solving with non-traditional algorithms. Artificial neural networks pertain to the family of connectionist techniques.

Habitually referred to simply as neural networks, Artificial Neural Networks (ANN) are a paradigm of automatic learning and processing inspired on the workings of the nervous system in animals. They consist of a simulation of properties observed in the biological neural systems, through mathematical models recreated by means of artificial mechanisms – like the integrated circuit of a computer. The idea is to get machines to give responses similar to those that a basic brain might give.

A neural network can be defined as a system for processing information that contains a great number of elements for processing (neurons) profusely connected amongst themselves, by means of channels of communication (Regueiro et al., 1995). These connections establish a hierarchical structure and permit interaction with objects from the real world. Unlike traditional computation based on predictable algorithms, neuronal computation can help us to develop systems that resolve complex problems whose mathematical formalization is hugely complicated (Guerrero Bote, 1997).

Artificial Neural Networks and Self-Organizing Maps

One variant of the ANNs is the Self-Organizing Map, or SOM. Its uniqueness lies in the fact that it works with bidimensional outputs (Moya-Anegón; Herrero-Solana; and Guerrero Bote, 1998). This is precisely one of the most important contributions of ANN in the field of domain visualization, developed by *Kohonen* (1985; 1997) and *Kaski* et al. (1998), who showed that input information alone presupposes an inherent structure, and so a functional description of the behavior of a network suffices in order to prompt the formation of topological maps.

What is most characteristic about this type of network is that it presents a competitive layer that classifies the training cells. The difference with other competitive layers is that here each neuron exerts a competitive influence upon itself as well as upon the topologically close or neighboring neurons.

The SOMs (Fig. 4.10) would constitute one of the most promising algorithms for the organization of vast volumes of information by virtue of their scalability, although they have a weak theoretical basis for the selection of learning parameters (Moya-Anegón; Herrero-Solana; and Guerrero Bote, 1998).

The first to utilize SOMs applied to the visualization of information was Xia Lin, who used them to obtain topological maps of documents (Lin; Soergel; and Marchionini, 1991; Lin, 1997). Some time afterward, they began to be used for automatically categorizing large amounts of documents (Chen; Schuffels; and Orwig, 1996; Chen et al., 1997; Orwig; Chen; and Nunamaker, 1997), for the visualization, search, and retrieval of information (Chen et al., 1998; Moya-Anegón; Herrero-Solana; and Guerrero Bote, 1998; Moya-Anegón et al., 1999; Guerrero Bote; Moya-Anegón; and Herrero Solana, 2002b), and the automatic extraction and analysis of terminological relations (Guerrero Bote; Moya-Anegón; and Herrero Solana, 2002a).

Despite being more difficult to interpret, the SOM product is much more appropriate than that of MDS or clustering for use in interactive environments (Lin; Soergel; and Marchionini, 1991), as it can be used as a tool for browsing (Chen et al., 1998). Moreover, it manages to make neighboring relations among groups more visible (Moya-Anegón and Herrero Solana, 1999).

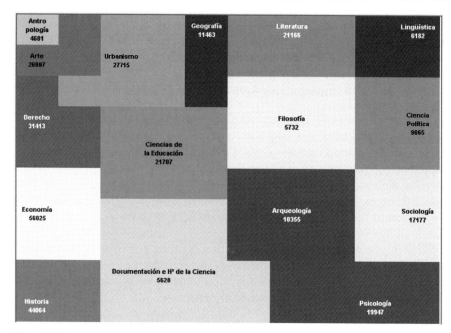

Fig. 4.10. Example of an SOM from NeuroISOC (Moya-Anegón et al., 1999)

4.4.3 Techniques Based on Social Networks

Social networks aspire to represent the behavior of units of analysis and the system as a whole by means of the interactions among component elements or nodes. Yet in most cases, the relationships among the units of analysis – nodes – form such a massive tangle of links that it is impossible to see the main relations. There is no easy solution to this problem; though several techniques and algorithms (pruning algorithms) have been developed to clear up the network by eliminating the least significant links. The result is a simplified network that, depending on the method behind it, represents more or less reliably, and more or less comprehensibly, the structure and essence of the original network.

Pathfindernetwork

The Pathfinder algorithm is a pruning algorithm developed in the core of cognitive science in order to determine the most relevant links in a network (Schvaneveldt, 1990). It extracts the principal structure of a network on the basis of the proximity of its variables. The result is a typical PFNET structure.

A social network and its links contain a number that indicates the distance between the nodes or the magnitude of the relation. This value can be used to prune away the less significant links. However, the pruning does not, by any longshot, afford a single solution, as the links that are not important for one given structure may indeed be significant for another.

The PFNET relies on two principal elements: the *Minkowski* distance, and an extension of triangular inequality.

The Minkowski distance, used to calculate the distance between two points over several links, is defined through a parametrical equation that subsumes the Euclidian distance for $r = 2$. This distance admits that r be stretched to infinitum, in which case it would be equivalent to finding the maximum of the intermediate distances. The parametrical equation of Minkowski is as follows.

$$D = \left(\sum_i d_i^r \right)^{1/r}$$

where r is the parameter associated to distance.

The second element to play a role in PFNET is the principle of triangular inequality (Fig. 4.11). This principle is based on the fact that one of the sides of a triangle can never be greater than the sum of the other two. In PFNET, the algorithm searches out the path between nodes that has the minimal weight or the least cost. The result is a graph with all the nodes connected by links that do not violate the principle of triangular inequality. That is, those paths whose weights are less than the sum of the combination of the weights of the other paths are the ones that remain. Distance, through the intermediate nodes, is calculated using the Minkowski equation.

The Pathfinder algorithm is defined by two parameters: r, associated with the *Minkowski* distance used; and q, related to length, understood as the number of links of the paths compared. Therefore, all the links that defy triangular inequality, having one associated distance that is greater than another for the same points of up to q links, and calculating the global distance of the second path using the Minkowski equation with parameter r, will be eliminated. The maximum value possible for q is *n-1*, where *n* is the number of nodes.

The networks resulting from analyses based on citation, cocitation or the co-occurrence of terms (authors, journals or ISI categories) tend to be

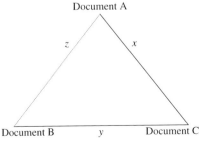

Fig. 4.11. Principle of triangular inequality. Links x and y are more important than link z

highly interconnected. For this reason the Pathfinder algorithm is most commonly used with $r = \infty$ and $q = n-1$.

The resulting network contains all the original nodes, but preserving only the most salient links. As can be seen comparing Figs. 4.12 and 4.13, the PFNET selectively prune links, eliminating a substantial proportion of the complexity and visual noise.

Although PFNET has been used in LIS since 1990 (Fowler and Dearhold, 1990), the first to put forth their utility in citation was *Chen*, who used hypertextual information to present a new means of organizing, visualizing, and accessing this information by what he denominated *Generalised Similarity Analysis* or GSA (Chen, 1998a, b).

Until today (within our field of knowledge), PFNET have been used for the study and representation of minor domains or scientific communities, described by a number of authors (Chen, 1998b; Chen, 1998a; Chen and Carr, 1999c; Chen and Carr, 1999b; Chen and Paul, 2001; Chen; Paul; and O'keefe, 2001; Chen and Kuljis, 2003; Kyvik, 2003; White; Buzydlowski; and Lin, 2000; White, 2001; Buzydlowski; White; and Lin, 2002; Lin; White; and Buzydlowski, 2003; White, 2003).

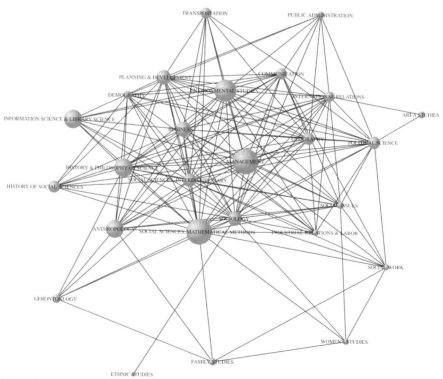

Fig. 4.12. Network of categories belonging to the Social Sciences, from the Spanish domain 1998–2002

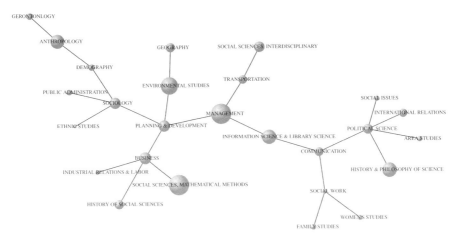

Fig. 4.13. Network of PFNET categories with values $q = n-1$ and $r = \infty$ belonging to the Social Sciences, from the Spanish domain 1998–2002

In achieving their domain visualizations, *White*, *Buzydloski* and *Lin* show a marked preference for PFNET. They affirm that both the SOM and the PFNET can include more elements in their displays than MDS or clustering can. And that, furthermore, SOM, MDS, and the dendrogram show similar elements as neighbours in space, whereas PFNET pairs them up by linking, indicating as well their degree of relation. Likewise, they argue that the groupings obtained with PFNET are easier to interpret by experts than those produced in any other type of display (White; Buzydlowski; and Lin, 2000).

For *Chen* and *Paul* (2001), PFNET is the best alternative, presiding over MDS, for the representation and reduction of space, as MDS does not reflect the relationships among the nodes. This makes it difficult to interpret the nature of each one of the dimensions represented. Yet PFNET explicitly incorporates the most significant connections, making the interpretation of the graphs depend on the links that connect all the elements within the graph, not the relative position of each dimension.

Budzydlowski compares the main techniques available today for reducing space and visualizing information. He believes that MDS lies halfway between SOM and PFNET: it resembles PFNET insofar as the methodology for showing information, for which reason he discards MDS and carries out a detailed study of the advantages and disadvantages of the SOM and PFNET. He then provides a thorough evaluation by a group of experts of online visualizations obtained through SOM versus those obtained similarly but using PFNET, for the process of information analysis and retrieval (Buzydlowski, 2002). The conclusions at the delta of this work are that SOM and PFNET are complementary techniques, offering similar results as far as the presentation of information is concerned, but PFNET has the upper hand in showing not only the positions of the nodes but also their most significant connections, besides being better for the generation of maps in real time.

Lin, *White* and *Buzydlowsky* (2003), as the result of their previous work, eventually disregard MDS and SOM as visualization elements and lean clearly toward PFNET as the optimal graphic interface in combination with ACA for the search and retrieval of information.

White adds one extra ingredient to domain visualization by using raw data in the matrixes. He upholds PFNET as the best guarantee for arriving at two-dimensional maps from raw data, with nothing but advantages over the rival techniques (White, 2003).

4.5 Spatial Distribution of Information

Just as the attributes or relations among variables can be interpreted through matrixes of distance or similarity measures, using MA techniques, these matrixes may in turn be represented by means of spatial distribution procedures, so that the distances/similarities between variables can be used to generate two- or three-dimensional maps where the similar variables appear together (and the different ones are separate from the rest). In most cases, the end result of these ordering techniques comes to form part of a graph or social network.

Basically, the core problem that has daunted researchers and analysts alike for over 25 years resides in the automatic generation of graphs: there is no blanket criterion as to how to make them. Two requirements are commonly admitted, however. One is to reduce, to the greatest extent possible, the number of crossed links, as *Moreno* (1953) advised 50 years ago. The other is to distribute the actors and the links in a uniform way over the network. Nonetheless, a severe application of the "first commandment" could interfere with the second: excessively reducing crossed links can affect the structure and uniformity of the network, which would impede its comprehension by the human viewer.

The automatic generation of graphs can be attempted in various ways, some of which entailing a limited positioning of the vertices. For example, the vertices may be placed in specific pre-determined coordinates (Batini; Nardelli; and Tamassia, 1986; Tamassia; Batista; and Batini, 1988), or they may form concentric circles (Carpano, 1980), or else parallel lines (Sugiyama; Tagawa; and Toda, 1981). Alternatively, the graph is interpreted as a physical system with energy stemming from its vertices, so that they are able to move about freely in the available space, thus reducing the system's energy and guaranteeing a good representation.

Of the many proposals for drawing graphs, we shall first briefly describe two of the spring embedder type, which are the ones most used in LIS and in domain visualization.

Spring embedders are algorithms dedicated to the representation and visualization of information. They get this name from the procedures upon which they depend to spread the information over the assigned space. Their main aim is to create attractive graphics, following a series of esthetic principles such as, for example, using the maximum space available, forcing the position of nodes, and reducing the number of crossed links. Thus we can obtain very successful results with graphs of a small size – some 50 nodes at the most. The spring embedders begin by assigning coordinates to the nodes, in such a way that the final graph will be esthetically agreeable to the human eye. This process is known as embedding (Eades, 1984) and it has been carefully studied by (Di Battista, 1998).

The two main extensions of the algorithm proposed by Eades have been developed by *Kamada* and *Kawai* (1989) and *Fruchterman* and *Reingold* (1991).

4.5.1 The Kamada–Kawai Algorithm

Kamada and *Kawai* propose an algorithm in which the position of the vertices is not restricted, and the links are drawn as straight lines. Thus, the purpose of the algorithm is to determine the position of the nodes or vertices alone.

The basic idea of the algorithm is as follows. It is believed that the desirable distance – be it geometric or Euclidian – between two nodes is the same as that represented in the graph, and so it is based on the *Floyd Warshall* algorithm (Gosper, 1998). A virtual dynamic system made up of rings or nodes, plus springs or links, is introduced so that the system will evolve, and at the same time the energy accumulated by the springs will diminish. Algorithmically, in order to avoid problems of computation, the evolution of each node is calculated separately, instead of summing up the total. That is, all the nodes are fixed except the one that accumulates the most energy; this one is allowed to evolve until the energy accumulated is lesser than a given limit, and then another is established. After this, again we choose the node that accumulates the most energy, which is permitted to evolve, and a limit is re-established. This process is repeated until none of the nodes within the network accumulates energy surpassing the threshold.

Very succinctly, this algorithm assigns coordinates to the nodes, trying to adjust as much as possible the distances existing among them with

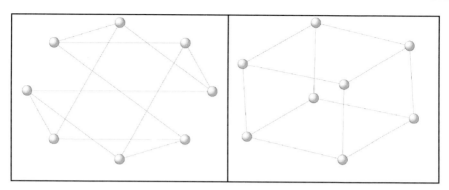

Fig. 4.14. Departing from the graph at the left, and applying the Kamada–Kawai algorithm, we obtain the graph at the right

respect to theoretical distances. Its use is widespread in the representation of social networks, owing to the assignment of a unitary distance to each link; and it offers very good results in the esthetic sense, along with adequate computation times for its application.

Figure 4.14 shows a typical example of the graph obtained using a *Kamada–Kawai* algorithm. If the algorithm is well programmed and fulfills all the principles set forth by its authors, its application to the graph at the left should give rise to the display at the right.

This algorithm has been profusely used in the field of domain visualization by authors such as *Chen, White, Lin,* and *Buzydlowski.*

4.5.2 The Fruchterman and Reingold Algorithm

Proposed for non-directed graphs whose links are drawn as straight lines, this algorithm produces drawings of graphs in two dimensions by means of simplified simulations of physical systems.

It is an algorithm for the placement of nodes based on the direction of forces. This method compares a graph to a mechanical collection of rings electrically charged (nodes) and connected by means of springs (links). The essential functionings are each two nodes reject each other, through repulsive force, while the adjacent nodes, which are those connected by a link, are mutually attracted through an attractive force. Over a series of iterations, the forces that model each one of the links are recalculated, and the nodes move around to reduce these stressful forces.

Figure 4.15 shows, at the right, the final graph obtained by applying the *Fruchterman* and *Reingold* (1991) algorithm to the graph at the left, which is the same one that appeared in Fig. 4.14, subjected to the *Kamada–Kawai* algorithm.

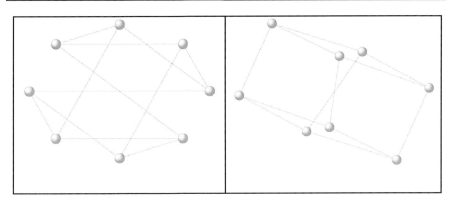

Fig. 4.15. Starting with the graph at the left, and applying the Fruchterman and Reingold algorithm, we obtain the graph at the right

The final result obtained either with the *Kamada–Kawai* algorithm or that of *Fruchterman & Reingold* is not very different: the two produce very similar structures, although the former shows superposition on two links, whereas the latter does so on three. Moreover, with the perspective used here to show the object and the place where the links cross over, the representation gained from the *Kamada–Kawai* algorithm is clearer and easier to grasp at a glance than the one reflected by the other algorithm. This is probably the reason why the *Kamada–Kawai* algorithm is the one most used for domain visualization.

4.5.3 Algorithm Evaluation

The criteria for evaluating this type of algorithm are fundamentally of an esthetic character. symmetry, the uniform distribution of nodes, the uniform length of the links, the reduction of the number of crossed links, all play a key role in the choice of one algorithm over another.

While studies such as that of *Brandenburg, Himsolt* and *Roher* (1995) do not detect any clearly predominant algorithm, most of the scientific community leans toward use of the *Kamada–Kawai* formula. The reasons behind this preference can be traced to its behavior in the face of local minima – the attempt to minimize the differences with respect to the theoretical distances throughout the graph, the good computation times, or the fact that it subsumes MDS when the latter uses the technique of *Kruskal* and *Wish* (1978). As indicated by *Cohen* (1997) and *Krempel* (1999), the *Kamada–Kawai* algorithm applies a criterion or energy that is similar to the stress of MDS as the measure of adaptation to the theoretical distances.

4.6 Information Visualization

Once the essential data of the variables has been obtained, together with spatial representations, it is necessary to generate displays that can be easily understood in an intuitive and efficient way, compatible with reality.

We have already explained how information visualization enhances the capability of interaction with large volumes of information, and aids researchers in detecting the underlying structures of their research fields. Visualization is essentially the design of a visual appearance for the variables and their relationships. A solid design should offer (Börner; Chen; and Boyack, 2003) the following:

- The capacity to represent vast amounts of information, on a large or small scale.
- Reduced time for the visual search of information.
- A good comprehension of the complex structures of data.
- The manifestation of relations that would otherwise not be perceived.
- A set of data observable from different perspectives.
- Encouragement as to the formulation of hypotheses.
- Provocation in the sense of analysis, debate, and discussion.

The achievement of all these requirements depends on the use of techniques such as filtering, panoramic views, blowing up and zooming in with the visualizations, their distortion for a better analysis of certain information, textual searches, and so on, and should be implemented one by one as independent tools for the analysis and visualization of domains. Notwithstanding, one may resort to commercial graphic formats or those of free distribution, which will allow for some of the above requirements.

There is no perfect choice of format overall for domain visualization. Rather, a wide variety exists: GIF, JPG, PostScript (PS), Encapsulated PostScript (EPS), Virtual Reality Modeling Language (VRML), or Scalable Vector Graphics (SVG), among others. In most cases, the choice is conditioned by the output format the researchers use; yet because it is vital to have quality images with a low weight in bits, so that they can be easily transported over the Web, recent years have seen a growing interest in Vector Graphics and their animation.

4.6.1 Scalable Vector Graphics (SVG)

There is no doubt about it: the vector graphic format that is best known and most used at present is Flash (Macromedia). Still, Flash may be criticized for being a compiled and extrinsic format, unlike Hypertext Markup Language (HTML) (W3C, 2003b) or Extensible Markup Language (XML) (W3C, 2003a), in addition to being an owner system subject to the needs and interests of a private firm. For this reason, in 1999 the World Wide Web Consortium (W3C) oversaw the development of Scalable Vector Graphics (SVG) (W3C, 2003c).

Specification SVG 1.1(W3C, 2004), published in January 2003 to override SVG specification 1.0 (W3C, 2001), offers the same graphic functions and animation functions as other owner systems, as well as added features. Because it is an XML application, SVG 1.1 has a perfect integration in the Web world, while at the same time important support on the part of the industry, since Adobe, Apple, Canon, Corel, Hewlett Packard, Macromedia, Microsoft, Kodak, and Sun, among others, have all contributed to the development of this specification (Jackson, 2002).

The SVG applications nowadays reach into fields as varied as mobile telephones, printing, web applications, aerospace graphic design, Geographical Information Systems (GIS), and so on. In its version 1.2, still in the developing stages, SVG is to include development of technology for SVG Basic and SVG Tiny cell phones, focused specifically on the design of low-consuming devices, as part of the 3GPP platform (the 3rd Generation Partnership Project), for building a new generation of mobile phones; the elaboration of directives for creating final print formats in XML (SVG Print), appropriate for the storage and printing of information; and a language for the interactive definition and presentation of SVG-specific tags: the sXBL (SVG's XML Binding Language).

4.6.2 Advantages of SVG as a Representation Format

Among the numerous advantages of working with SVG, the main differences with respect to its competitors are as follows:

- It uses XML as a descriptive language. The manufacturers of software are beginning to use this language as a native format in their applications. Therefore, SVG will be compatible with any application that recognizes XML.
- The file is very compact in size, less than that of its equivalents. Codified in maps of bits, it allows us to represent three types of objects in one same graph: vector graphics, images, and text.
- It features all the pros of the vector format: it is scalable, compact, and its forms can always be edited through Bézier curves, with softened and transparent outlines.

- The graphic objects can be designed, grouped, transformed, or combined with other objects included in the format, and it can even embed maps of bits.
- The text is editable: it admits the most common scalable fonts, such as TrueType or Type 1. This is a significant difference from the present-day GIF or JPG: the text that they contain can be edited, selected, searched, indexed (search engines). The high quality of the graphics does not depend on their resolution, and size can be increased or reduced using a zoom without sacrificing quality.
- The quality of the color is excellent and the tones of the graphics can be calibrated with standard color processing systems.
- Although it is particularly convenient for the design of two-dimensional graphs, it can also be used to simulate three-dimensional ones.
- The graphs can be generated in a dynamic form in a Web server as the response to instructions in Java, JavaScript, Perl, ASP, PHP, and so on.

5 Material Used

The elaboration of this study was founded on the information gathered and stored in the Web of Knowledge (The Thomson Corporation, 2005b) through their different databases. The reasons for using these data sources are explained below.

The ISI products[1] have a very different structure than most commercial databases, as they include the references or bibliographic citations upon which the authors support or justify their claims.

Insofar as thematic representativity, we should point out that this source gives excellent results in view of the goals of our study. The ideal would be to attain total coverage of a discipline. Yet these databases, being multidisciplinary sources, are the most homogeneous referent available at present for creating balanced representations.

One of the limitations of these databases, meanwhile, is the multicategorization of their documents. The JCR assigns one or more subject categories to each of the journals downloaded from their databases, determined by the subject matter that each journal itself declares to cover. This implies that any work published in a journal with multithematic assignment will automatically acquire that same peculiar status, which brings about a multiplying effect in category cocitation: these will be cocited as many times as the number of categories to which the cocited document is assigned. The end consequence is that all the categories, to some extent, appear cocited amongst themselves, and thus we lose one of the characteristics most essential for bibliometric distributions: the degree of the nodes is unequal to the right, a phenomenon known as *power-law* (Chen and Hicks, 2004). The graphic representations of these cocitation matrixes may be considered as networks of the small-world type (Watts and Strogatz, 1998).

Another limitation has to do with documental coverage. ISI databases index a tremendous number of journals, both national and international, but they do not present equitative percentages for each of the different countries, nor do they cover monographs or reports. Such limitations affect the humanities and social sciences more than the hard sciences, and the applied sciences more than the basic sciences. In "peripheral" countries, the use of these databases does not convince the scientific community on

---

[1]Actually copyrighted as *Thomson Scientific*

the whole, as many experts believe that it penalizes researchers who publish in languages other than English, just as it overlooks those with lines of research that hold local or regional interest but could hardly find a niche in journals with greater impact factors (García-Guinea J. and Ruis J.D., 1998). Indeed, there are scientists who question the selection of specialized journals carried out the by the ISI on the basis that they are the reflection merely of basic research, not applied research (Sanz E.; Aragón I.; and Méndez, 1995). Therefore, we should acknowledge the fact that the limitations of these databases reside mainly in the Social Sciences and the Humanities, given their more regional characters, causing their output to escape the control of international databases to a greater extent (Kyvik, 2003). That is, national publications have more relevance in linguistic and cultural fields than in the non-culturally determined areas of science and technology (Bordons and Gómez Caridad, 1997).

The history of the ISI databases has been plagued with criticism of bias in journal coverage, both in terms of disciplinarity and nationality; yet recent studies (Braun; Glanzel; and Schubert, 2000) that compare SCI coverage with that of Ulrich's International Periodicals Directory (U-S&T), show this to be false. The bulk of SCI journals present a fair balance with respect to the U-S&T at the macrolevel, with regard to countries and disciplines, at least. Contrary to public opinion, the cited study demonstrates that there is no bias in these databases in favor of the United States or of Biomedicine (which may, in fact, be underrepresented). Exceptions concerning coverage by disciplines are centered on Germany, and more specifically on Agriculture; while regarding editors, Elsevier stands out (Moya-Anegón et al., 2004). In general, there is an overrepresentation of the main publishing companies in the SCI, but at any rate this does not interfere with the objectives of the present study.

For the specific case of science in Spain, the selection of this source is coherent with current legislation[2] establishing the criteria for evaluation of Spanish research in all the scientific fields except Law and Jurisprudence, History, Art, Philosophy, Philology, and Linguistics. This has led Spain's scientists to direct their articles to scientific journals in the ISI databases, which are the ones indicated by the National Evaluating Commission (*Comisión Nacional Evaluadora* or CNEAI) as the key of reference in assessing research for the concession of incentives (Jiménez Contreras; Moya-Anegón; and Delgado López-Cózar, 2003).

All in all, despite its gaps and weak points, the ISI databases stand today as the best tool for obtaining data that can be handled with guaranteed reliability. This is the number one reason they are used by all the governments and institutions of the western world as the point of reference, and are used

---

[2] Resolution from 08/28/1989, modified and completed by R.D 1325/2002

worldwide to a considerable extent in evaluating research activity. We hold them to be an adequate source of information, in that the data they supply consistently reflect world research with international visibility, and therefore can be applied to arrive at the graphic representation and structural analysis of worldwide scientific production.

5.1 Data Extraction

For strictly investigative purposes, on the August 2, 2004, we finished downloading from the Web of Science (The Thomson Corporation, 2005b) and more specifically from the Science Citation Index-Expanded (SCI-EXPANDED), the SSCI, and the A&HCI, all those records of scientific documents published worldwide in the year 2002. That is, all those that in the database field "Year" contained the string of characters corresponding to 2002. Likewise, we downloaded all the records that in the field "Address" contained the word Spain, and in the field "Year" showed a year from 1990 to 2002 (both included).

The Journal Citation Report, or JCR (The Thomson Corporation, 2005b), assigns each journal to one or more thematic categories. In order to reassign a subject matter – ISI category – to each document, we extracted this information from the JCR, in both its Sciences edition and its Social Sciences edition for 2002.

Then, after data extraction, we carried out a series of tasks related with data quality control. They are explained just below.

5.2 Data Analysis and Treatment

The ISI databases, due to their open period of coverage (1945 to the present for the SCI; 1956 to the present for the SSCI; and 1975 to the present for the A&CHI), their evolution over time and the different formats and media that they have involved, present a series of problematic circumstances for processing, which could be quite easily overcome in some cases, and required well-designed solutions in others.

5.2.1 Normalization of Journal Titles

The first problem we came across was the inconsistency of some abbreviated journal titles in the JCR listing, and of the presentation of references to the cited articles. This led to a loss of information when working with citation and cocitation analysis, something we must assume from the start.

In our case, out of the 7,590 journals included in the JCR 2002, only 98 of the abbreviated titles with at least one citation had no correspondence with the form introduced in the ISI databases. That is, just 1.29% of the total of abbreviated titles, with at least one citation, do not coincide with the titles listed in the JCR. For the case of the Spanish domain 1990–2002, the proportion is slightly less. We believe that any potential bias from such a low percentage is negligible.

5.2.2 Multidisciplinarity of Journal Contents

Another point to bear in mind with the ISI databases is the assignment of the category Multidisciplinary Sciences to a certain group of journals embracing a broad range of interests, such as *Science, Nature, Endeavor* or *Interscience*, to name a few. This, while logical, creates unexpected consequences down the road: Works that are dealing with a specific discipline or category, say for instance genetics, would appear distant from the rest of the studies concerning genetics, owing to their publication in multidisciplinary journals, and therefore being tagged as such and only as such.

This problem is not easy to solve. The European Commission, in its Third European Report on Scientific and Technological Indicators (European Commission, 2003) directly eliminates the category Multidisciplinary Sciences, along with the indicators deriving from the documents under this heading.

In our study, in order to avoid losses of information of this sort, we substitute the category of the documents assigned to Multidisciplinary Sciences, using instead the category most cited by the references of each one of these documents. In the cases where the category Multidisciplinary Sciences happens to be the most cited one, the document is assigned to the second most cited category. But when the only category cited is Multidisciplinary Sciences, and we have no other references, we retrieved each one of these source documents, and by studying the title and abstract, we manually assigned it to a JCR category, in view of contents.

5.2.3 Author Control

The ISI does not perform any sort of authority control in its databases. This leads to the appearance of different entries for a single author, and the solution to this problem calls for a great deal of time and effort. Nonetheless, this drawback does not hamper the study or distort the results, as we work at higher levels of aggregation where the author names are not actually used in any step of processing.

A similar problem resides in the assignment of a country of origin for the document. The information that authors give as to their citizenship is directly transferred, without any reconfirmation or comparison, to the field "Country" in the database. Thus, for example, we do not find countries such as United Kingdom, whereas we may encounter others such as "Gibraltar".

This final difficulty arises mainly at the time of document selection in a specific geographic area. To avoid it we have no other choice but to corroborate if the geographic unit or domain we hope to analyze actually corresponds with the country name as registered in the databases (see Annex I); if this is not the case, we need to combine the geographical units that the ISI considers as countries, until constituting the appropriate geographic unit.

5.3 Generation of the Secondary Source

Once these trouble spots related with the consistency of the data in their original state have been detected and resolved, all the information is transferred to an ad hoc relational database.

The final result is a repository holding the structured information from the documents and the relationships established a priori. In the case of the world as a whole, with its 1,751,996 different scientific authors from 206 countries – Annex I – who have contributed (on their own or in collective efforts) have published the total of 901,493 documents (articles, biographical items, book reviews, corrections, editorial materials, letters, meeting abstracts, news items, and reviews) gathered in 7,585 journals accepted by the JCR and to which the 219 categories acknowledged by the JCR for the year 2002 (Annex I). With the exception of Multidisciplinary Sciences, where the procedure was necessarily altered as explained in Sect. 5.2.2., the total number of citations made by the 901,493 documents and registered in this database came to 25,682,754.

For the evolutionary or temporal study of the Spanish domain, we counted 1,095,210 authors who, alone or in collaboration, saw the publication of their 294,778 documents in 10,404 different journals accepted by the JCR. These covered 240 categories listed under the JCR from the year 1990 to 2002 (Annex I), with the exception, again, of Multidisciplinary Sciences. The total number of citations recorded in this new database was 7,364,747.

6 Methodology

The methodology for creating offline representations has not changed much since the 1980s: the area of knowledge to be studied is selected, the units of analysis are chosen (works, documents, an author's works, journals, country of publication, etc.), and the data are collected and transferred to a co-occurrence matrix that is later transformed into a proximity matrix. Visualizations are obtained using clustering, FA and/or MDS over co-occurrence matrixes, say, by authors; in such a way that the hierarchical structure of the clusters produces groupings and the MDS ordering of the items in an n-dimensional space (White and McCain, 1997).

Just 8 years have gone by since White and McCain made this statement, and from then till now new methodological contributions have come to improve those already in existence. With time, more will appear on the horizon. Ours is presented below.

6.1 Category Cocitation

To date, the cocitation of classes and categories (Moya-Anegón et al., 2004), together with the combination of clustering and MDS (Small and Garfield, 1985) have been the only two models proposed for the representation of large domains.

In the present study, we adopt class and category cocitation as the model behind the display. We also propose or introduce some modifications, however.

We consider that JCR categories alone constitute informative units sufficiently explicit so as to be used in the representation of the different disciplines that make up science in general. These categories, in combination with adequate techniques for the reduction of space and the display of information, allow us to build scientograms of science on the whole – or of other vast domains – that will prove clearer, much more informative, and more user friendly, especially for non-experts, than those obtained though class cocitation, because they will facilitate comprehension and handling. Furthermore, the fragmented representation of a domain (to which ANEP class use is funneled) makes it impossible to study the connections among categories of different classes, because they are not represented. Such considerations led us to use the amplified model of cocitation introduced by *Moya-Anegón* et al., (2004), but only as far as the level of categories. Likewise, we set forth some considerations and modifications that, in our opinion, improve the method.

The model of category cocitation, as described by its authors, suffers from two problems that have slowly become evident over time and usage, as this technique has been extended to all sorts of domains.

1. The first problem is in reality an implicit defect, which appears as a consequence of the adoption of JCR categories as units of measure and cocitation.
2. The second problem is related with the standardization of the cocitation value of the different databases used to build the data source and the need to offer raw data cocitation values. Below we will describe these while proposing a solution.

6.1.1 Latent Cocitation

The scheme of cocitation we adopted (Fig. 6.1) and the intrinsic idiosyncrasies of the JCR classification can, admittedly, lead to errors of accumulation in the computation of category cocitation. This error is the consequence of what we call latent cocitation. In sum, a single citation of a document that has been published in a journal assigned to several JCR categories will trigger the automatic cocitation of all these categories, though they have not actually been cited by different documents. To prove our point, and going back to Fig. 6.1 and the Journal of the American Society of Information Science and Technology (JASIST): the reference by a given document X of another that has been published in this journal will automatically unchain the cocitation of both JCR categories, to which the journal has been ascribed. That is, a single reference to a JASIST article will mean, unintentionally, the cocitation of the category Computer Science and Information system as well as that of Information Science and Library Science, with the ensuing overestimation of cocitation or mutual relationship between the two categories. This disembarks in an error in the representation of domain structure.

Eliminating such latency in cocitation can prove easy. We need only to group the categories cited by each one of the source documents, and then, over that grouping per se, to carry out cocitation calculations. The result is a non-latent cocitation of categories, where latency inherent to classification does not affect the representation of the domain to be analyzed.

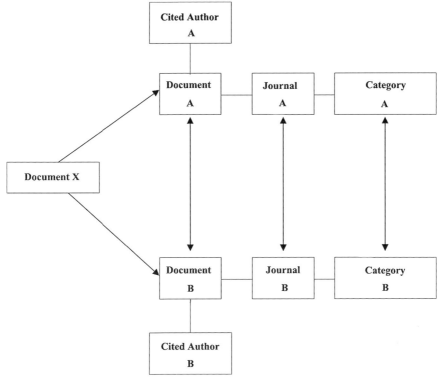

Fig. 6.1. Model for category cocitation (*adapted from* Moya-Anegón et al., 2004)

This non-latent cocitation will be the type we use for the generation of the scientograms of vast domains presented in this study; and heretofore, when we speak of cocitation, we will be referring to non-latent cocitation.

6.1.2 Raw Cocitation Values

The calculation of cocitation produces a matrix of values. We shall work with matrixes of category cocitation, and all of them will be square or symmetric. That is, they will have an equal number of lines and columns, and so the coincidence of their elements or categories will always be recip-rocal, not directional.

In order to generate the category cocitation matrix, we toss out a query of crossed references, to be executed against the relational database or sec-ondary source.

The result is a symmetrical matrix, of N by N categories, where N is the number of categories existing in the output per domain. We are not interested in representing nor in analyzing the auto-cocitation of cate-gories, for which reason we place zero values in the diagonal of the matrix, thereby eliminating the cycles or loops in each category.

Like *White* (2003), we have seen that the introduction of standardiza-tion measures in the values of the matrix – be they Pearson, the cosine function (Salton; Allan; and Buckley, 1994) or the *Salton* standardization measure (Salton and Bergmark, 1979) – consistently produces distortion in the distribution of the map information. After much testing, we con-clude that when using Network Analysis tools, the best visualizations are those obtained with raw cocitation data as the unit of measure. Notwith-standing, and in view of the need to reduce the number of coincident cocitations so as to improve the yield of the pruning algorithm, to the values of raw data cocitation obtained, we add the corresponding value of cocitation after standardization. In this way, we manage to work with raw cocitation values while differentiating the valued similarity measures among categories with equal cocitation frequencies. We achieve this by introducing a slight modification into the standardizing equation pro-posed by *Salton and Bergmark*:

$$\text{SCM}(\ ij) = Cc(ij) + \frac{Cc(ij)}{\sqrt{c(i) \cdot c(j)}} \qquad (6.1)$$

where

SCM is the standardized cocitation measure
Cc is cocitation
c is citation
i, j are categories.

These cocitation matrixes are the basis and origin of the scientograms that make manifest the structure of the domain represented, along with the rela-tions or interchange of knowledge that takes place between the different categories or disciplines of a scientific domain.

Here we show the raw cocitation data values, already normalized, obtained by applying the above formula to nine JCR categories in a fictitious domain, to arrive at the corresponding degree of cocitation (Table 6.1).

6.2 Dimensional Reduction: The Basic Structure of a Domain

The input of data over the past twenty-some years has involved raw data cocitation, or else standardized values. Techniques used for the reduction include MDS, clustering, FA, SOM, and PFNET.

Table 6.1. Example of a cocitation matrix of 9 by 9 categories

	Energy & Fuels	Geosciences, Interdisciplinary	Engineering, Petroleum	Gastroenterology & Hepatology	Radiology, Nuclear Medicine & Medical Imaging	Mathematics	Medicine, General & Internal	Education, Scientific Disciplines	Medical Informatics
Energy & Fuels	0	26.31	30.2	0	0	0	2.09	0	0
Geosciences, Interdisciplinary	26.31	0	16.1	0	3.02	2.07	2.3.4	1.1	0
Engineering, Petroleum	30.2	16.1	0	0	0	0	0	0	0
Gastroenterology & Hepatology	0	0	0	0	36.03	0	248.1	0	1.01
Radiology, Nuclear Medicine & Medical Imaging	0	3.02	0	36.03	0	0	90.04	1.07	7.02
Mathematics	0	2.07	0	0	0	0	0	2	2.01
Medicine, General & Internal	2.09	2.3.4	0	248.1	90.04	0	0	5.01	24.3
Education, Scientific Disciplines	0	1.1	0	0	1.07	2	5.01	0	0
Medical Informatics	0	0	0	1.01	7.02	2.01	24.03	0	0

For the work presented here, PFNET loomed as the option of choice. It appeared to be best for preserving the most significant semantic relations while still capturing, in an economic fashion, the intellectual structure of a domain from a social standpoint. After all, semantic relations and their extents are established by the very authors who make up a given domain.

For the first time, then, we face the feat of displaying on a plane – whether the computer screen or a sheet of paper – the structure of scientific output of vast geographical domains. And all this without bending under the influence of geopolitical connotations: the domain may be a state, a country large or small, even a continent. And why not the world?

The adoption of JCR categories as cocitation units also implies accepting them as units of representation. This meant the subsequent scientograms having, as a rule, over 200 categories. The display in two-dimensional graphics of such a high number of elements along with their corresponding relationships, in such a way that the human eye can easily interpret the information, was a true challenge. We looked to Small for guidance:

> For reasons of clarity in visualization, it is better to eliminate some connections... The loss of information in the structure implies a gain in simplicity, for which reason the sacrifice is, in some cases, justified (Small, 2000).

And also to *Hjørland and Albrechtsen*

> If we are dealing with a system with excessive possibilities, in which no priority has been given to the essential connections, we will only arrive at an excess of information on the part of the user, and a lack of effectivity as regards the system (Hjørland and Albrechtsen, 1995).

The PFNET lends solutions in these directions, and is very useful for shedding light on the scientific and intellectual structure of vast scientific domains.

We should underline that we are working with cocitation values, which in reality are nothing more than the values of the links; and that these values range from one to some thousands, in the best of cases. Yet, as occurs in all bibliometric distributions, the low cocitation values are the most abundant and most likely to be repeated. This implies that the greater the number of nodes, the greater the possibility that the value of the links connected to a single node be the same. As the system for pruning links used in the PFNET networks is based on the principle of triangular inequality, it may occur that the algorithm cannot eliminate some of the links of a specific node, because it finds two paths with the same distance. The materialization of this probability gives rise to the appearance of loops in the network, and thus increased density, detracting from the clarity of visualization and comprehension of the matter.

And so, one added value with the use of the modified *Salton* and *Bergmark* formula is that it impedes the appearance of loops or cyclical connections among nodes in the domain representations. This modification allows us to practically eliminate the probability that the values of the links connecting a single node are the same, and thus that loops appear in the scientograms. This is because the integer numbers of the unstandardized cocitation value become real, and trail as many decimal numbers as deemed necessary, in order to break up the loops.

To illustrate what we have explained up until this point, we present and comment on three PFNET with values $r = \infty$ and $q = n - 1$, as seen in Figs. 6.2, 6.3 and 6.4. They were built from a single cocitation matrix of 215 by 215 categories, and produce three distinctive visualizations of the same fictitious domain. In the first, the cocitation values were standardized with the *Salton and Bergmark* formula per se. The second used the raw data cocitation values. Finally, the third uses the cocitation values obtained with the modified method explained above.

The PFNET of Fig. 6.2 was built on a schematic and user-friendly graph in order to facilitate comprehension for non-experts. It has a low density in links and connections between nodes, and stretches out over space. However, its depiction of the scientific structure of the domain is quite linear. With the exception of the low central part, marked in a darker tone, where a node seems to constitute the point of confluence of other nodes, this network does not present any clearly differentiating element that might be considered as the point of departure for analyzing the network structure. The truth is, if it were not for that red point of confluence, indicating a strong interchange of information, the structural analysis of this network might begin with either of its distant end points, equally valid.

The PFNET of Fig. 6.3 presents a complex graph, whose brambles are difficult to interpret. This type of representation is typical of those obtained when the raw cocitation value is used. As we mentioned

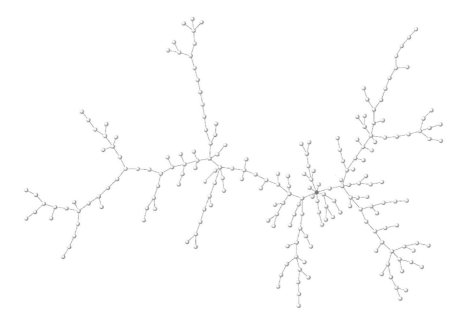

Fig. 6.2. PFNET of category cocitation with values standardized using the Salton and Bergmark formula

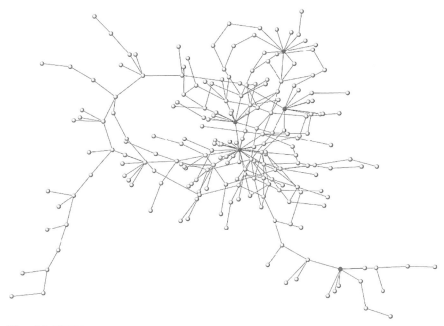

Fig. 6.3. PFNET of category cocitation with pure cocitation values

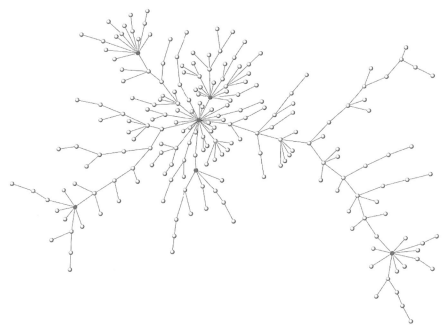

Fig. 6.4. PFNET of category cocitation with values standardized according to our proposal

previously, it is precisely the use of this unstandardized value that makes a pruning algorithm unable, at times, to locate the longest path (to be eliminated), thus inducing the appearance of loops between nodes. Yet by using raw category cocitation data, we give rise to scientific groupings surrounding different nodes – those marked in a darker tone. Despite the high degree of interlinking in this network, the red nodes are easily detected and can serve as focal points for analysis of the scientific structure of the domain.

The PFNET network of Fig. 6.4 is schematic, clear, and user friendly, both for viewing and for handling. In itself, it identifies (no need for technical aid) the darker nuclei of the main areas of research in the domain represented. These areas, clearly visible, and the number of nodes that make them up reflect the different grades of relevance of the elements of the structure and its analysis as a whole.

This third type of graph condenses the virtues of the other two, combining simplicity with the automatic identification of the major areas of knowledge of the domain. This obeys the logic of uniting the best ingredients of two proposals, and leaving their weak points behind.

6.3 Spatial Distribution of Information

Of the most popular algorithms for producing graphic representations of the PFNET-type networks, we opted for that of *Kamada and Kawai*, coinciding in our choice with the majority of the international scientific community. Our reasons are as follows:

1. It is a very simple algorithm, yet adequate in automatically generated social networks with a very low computational cost.
2. It gives good results in graphs that contain hundreds of vertices, in a reasonable period of time.
3. It spreads the nodes over space in such a way that the most similar ones tend to appear together, whereas the dissimilar ones are farther apart.
4. It can incorporate different variations.

We have elaborated (with "extra" information) some of the elements appearing in the scientograms. This required our introducing minor variations in the algorithm in order to avoid the overlapping of nodes and/or links.

The first variation has to do with the size of the nodes, to make the PFNET scientograms more informative. We make each node correspond in proportional size to the totality of the scientific output of the domain represented.

The second variation consists of showing the thickness of the links. As in the previous modification, here we also pursue improved informative capacity of the scientograms. In our opinion, this complements to the pruning algorithm: thanks to the principle of triangular inequality, we reduce the number of links of the network to just the most significant connections; and with this modification, we show which of these relationships are the strongest. In other words, the categories with a greater degree of interaction are made apparent by the number and thickness of their links.

The inherent advantages of the algorithm selected, along with the proposed modifications, will bring us alongside our objective of creating clear, schematic, intelligible, and attractive representations for human understanding. A quick comparison of the scientograms of Figs. 6.5 and 6.6, made using the same PFNET values, renders obvious the reasoning behind choosing the *Kamada–Kawai* algorithm in lieu of the *Fruchterman and Reingold* algorithm.

Both visualizations occupy all the available space, spreading out their nodes and links according to the principles established by each algorithm, though one is more successful than the other. Both include variations for the thickness of the links and the size of the nodes. Both, too, gain in informative capacity by making the size of the nodes proportional to the volume of output in each category, though perhaps at the expense of some loss of clarity in the display if we compare them, for example, with the network of Fig. 6.4, where the sizes are not reflected.

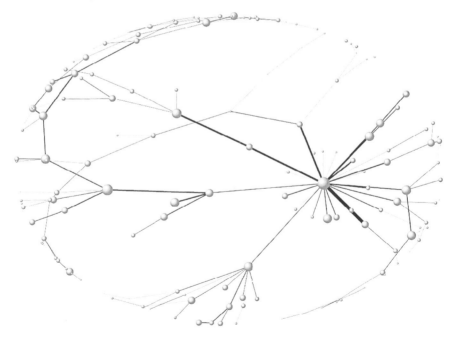

Fig. 6.5. PFNET made with the Fruchterman–Reingold algorithm

Fig. 6.6. PFNET made with the Kamada–Kawai algorithm

The Fig. 6.5 presents a radial scheme, where all the nodes most related with the rest are in central positions, while those with fewer links appear in the periphery. This structure is typical of those obtained using MDS. In cases such as this, where just a few nodes have a protagonistic connection with the rest, we end up with representations that are practically empty in the middle, with a heavy agglomeration in the periphery. If, moreover, we add a high number of nodes (as would be the case in our domain), the peripheral display would be difficult to discern, owing to the lack of space in which to place so many categories. The result would be an overlapping of nodes and links.

Figure 6.6 shows an extended, user-friendly, and very informative network, without any overlapping links or nodes.

The result obtained with the *Kamada–Kawai* algorithm is something as spectacular and visually informative as a railroad or underground metro map.

- One glance suffices to locate the center and the limits of the domain, for orientation.
- It is easy to jump from one category to another by following the links, just as train stations are joined by lines of rails.

- One can see which "stations" are the most important in terms of the number of connections they have, and which act as intermediaries or "hubs" with other lines, or as forking points.

The *Kamada–Kawai* algorithm, in conjunction with the intrinsic characteristics of PFNET and the use of raw cocitation values, affords two unique advantages (White, 2001), that are also seen in conjunction with category cocitation:

1. It makes the main scientific groupings bloom into a depiction of the domain structure.
2. It allows one to visualize how these groupings are chained together in explicit sequences, where the order of the nodes in sequences is not arbitrary, but rather reveals the way in which the major scientific groupings of a domain are interconnected, thus lending logical order to the underlying structure.

These two characteristics, when combined with FA — to which we referred in the previous section — facilitate a quick comprehension as well as the deep analysis of a domain.

6.4 Methodological Considerations

We now proceed to spell out some particular properties of our scientograms of large scientific domains, in light of the data and methodology applied.

6.4.1 Weak Links

As stated in Chap. 5, one of the problems deriving from the use of the ISI databases is the incomplete documental coverage: a considerable number of journals from different countries, and documental types such as monographs and reports, may be overlooked.

From the standpoint of information visualization and for the concerns of the work we present, this lack of coverage makes cocitation between certain JCR categories surprisingly weak in some instances and oddly strong in others. As a result, when PFNET is applied as the pruning algorithm, connections put in touch categories with very little inter–relationship, due to tangential cocitation, yet prevailing over others that are more natural but with weaker connections. For example, in the case of Fig. 6.7, the category Philosophy appears linked to the category General and Internal Medicine, toward the middle-right of the scientogram.

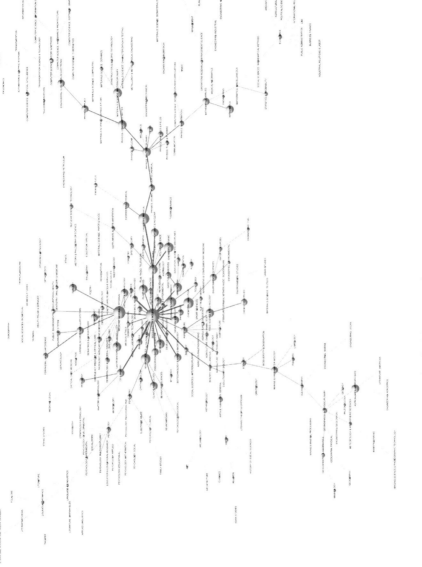

Fig. 6.7. Scientogram of PFNET of a fictitious domain

Although this connection between disciplines is by no means bizarre, it comes as a bit of a surprise in the context of the mental models we may have assumed a priori regarding the interrelations of science on the whole; especially if we are trying to unearth the most relevant connections. If we look at this connection more closely, we see that Philosophy appears cocited with 28 other categories, including itself, as many times as the number appearing to the right in Table 6.2 would indicate.

The highest cocitation of Philosophy is with its own category. Yet as we said under Sect. 6.1.2., the value of the diagonal of the matrix, standing for self-cocitation, would be set at zero. Thus discarding the category itself as a unit of cocitation, we observe that the next most cited category for Philosophy is General and Internal Medicine, with a value equal to three. The other categories follow with values between two and one.

We know that the interdisciplinarity of science means that any type of connection is plausible. In this case what is strange is that some very akin categories, say History, do not appear cocited with Philosophy; whereas others such as History & Philosophy of Science have a cocitation value lower than General and Internal Medicine. This leads the PFNET, when it comes into action, to look for the shortest paths in order to prune them, selecting the one with the highest value. The result would then be the connection Philosophy–General and Internal Medicine. This confusion, again, is caused by the lack of documental coverage of the databases used, augmented perhaps by the short time period of the data used here – the year 2002 – along with the policy for author citation of certain geographical areas.

This obstacle can be easily overcome by using journal citation as an element of thematic assignment. That is, finding out which is the most cited journal for each category that has strange or unexpected connections, and linking it manually to its category, although it could also be solved by eliminating them and their links.

Beyond what is strictly necessary we prefer not to manipulate the construction of the scientograms. So we leave the connections just as PFNET establishes them. However, we would like to warn the viewers of our scientograms that most connections between categories can be considered trustworthy, while others should be taken with a grain of salt.

After observation of hundreds of PFNET scientograms of large domains, and the one-by-one study of their connections, we arrive at two conclusions. The first is that a link between two categories will be less dubious the greater its cocitation value. And the second is that the problem of coverage and the ensuing errors of connection occur above all in categories belonging to the databases of the Social sciences, particularly Arts and Humanities.

For this study and the time period it covers, we suspect that the links with a value under four are responsible for any dubious connections among categories. To make the reader more aware of these questionable links, we will show them in a softer tone.

Table 6.2. Categories cocited with Philosophy

Category	Category	No. of cocitations
Philosophy	Philosophy	26
Philosophy	Medicine, General & Internal	3
Philosophy	History & Philosophy Of Science	2
Philosophy	Ethics	2
Philosophy	Pharmacology & Pharmacy	2
Philosophy	Chemistry, Medicinal	2
Philosophy	Biochemical Research Methods	2
Philosophy	Psychology	2
Philosophy	Engineering	2
Philosophy	Mathematics, Applied	2
Philosophy	Chemistry, Analytical	2
Philosophy	Education & Educational Research	1
Philosophy	Environmental Studies	1
Philosophy	Ecology	1
Philosophy	Mathematics	1
Philosophy	Computer Science, Artificial Intelligence	1
Philosophy	Biology	1
Philosophy	Economics	1
Philosophy	History	1
Philosophy	Infectious Diseases	1
Philosophy	Zoology	1
Philosophy	Medicine, Legal	1
Philosophy	Pediatrics	1
Philosophy	Physics, Multidisciplinary	1
Philosophy	Planning & Development	1
Philosophy	Public, Environmental & Occupational Health	1
Philosophy	Social Studies	1
Philosophy	Social Sciences, Biomedical	1
Philosophy	Immunology	1

6.4.2 Isolated Nodes

In a PFNET, what is most important is not the exact placement of the nodes, but their interlinking (Buzydlowski, 2002). We would like to extend this statement: in a PFNET, what is most important of all is the existence or absence of links among nodes, since it is the links that supply information about the nodes and determine their positioning.

Just as the connection between two categories evidences the most significant or strongest relationship there among, and makes the position of the two one of adjacent placement, the absence of connections in a category indicates that it has no intellectual interchange with other categories, and therefore its position is not adjacent to any other one. Instead, it is located freely in the available space of the network. In sum, the absence of links can be just as important and informative as the existence of links.

It is not usual to find an isolated node on a scientogram of categories, but it is a possibility. This would happen exclusively when a category appears cocited only with itself, with no links to any other category. Such a finding may be determined by the scientific community in view of their cociting trends, a social factor to be taken into account analytically; or else it might be due to deficient documental coverage, causing us to lack information regarding cocitation with other categories. As we shall see later on, in Chap. 7, the categories of Social Sciences, Arts and Humanities are again the ones that tend to appear in isolation, essentially owing to the lack of coverage.

6.4.3 Information Retrieval

The offline visualization of static representations can be made more interesting by orienting the construction of interface prototypes toward information retrieval (White and McCain, 1997).

In order to enhance information retrieval, each PFNET scientogram includes hyperlinks in its links and categories. These hyperlinks allow Web consultation of the secondary source.

There are two ways to access and retrieve information. The first is category-dependent. It consists of a simple query that, for each scientogram, shows the work belonging to each category ordered according to document type (Fig. 6.8). The second shows the documents associated with the links existing between two categories, also ordered by document type (Fig. 6.9).

Due to limitations of the software used in the relational database, we are only able to show in complete form that information concerned with the domain Spain. For other geographic domains, the most detailed information we can offer is that referring to the journals where the documents have been published.

SPAIN - INFORMATION SCIENCE & LIBRARY SCIENCE - DOCUMENT LIST

Title	Journal	Year	N°	Vol	Pag	Type of doc.
Document organization using Kohonen's algorithm	INFORM PROCESS MANAG	2002	38	1	79-89	Article
An algorithm for term conflation based on tree structures	J AM SOC INF SCI TEC	2002	53	3	199-208	Article
A context vector model for information retrieval	J AM SOC INF SCI TEC	2002	53	3	236-249	Article
Advantages and limitations in the use of impact factor measures for the assessment of research performance in a peripheral country	SCIENTOMETRICS	2002	53	2	195-206	Article
Multivariate evaluation of Spanish educational research journals	SCIENTOMETRICS	2002	55	1	87-102	Article
Research productivity of scientists in consolidated vs. non-consolidated teams. The case of Spanish university geologists	SCIENTOMETRICS	2002	55	1	137-156	Article
Methods for the analysis of the uses of scientific information: The case of the University of Extremadura (1996-7)	LIBRI	2002	52	2	99-109	Article
Creating e-books in a distributed and collaborative way	ELECTRON LIBR	2002	20	4	288-295	Article
Participative knowledge production of learning objects for e-books	ELECTRON LIBR	2002	20	4	296-305	Article
Obtaining feedback for indexing from highlighted text	ELECTRON LIBR	2002	20	4	306-313	Article
A system to generate electronic books on programming exercises	ELECTRON LIBR	2002	20	4	314-321	Article
EIS evolution in large Spanish businesses	INFORM MANAGE-AMSTER	2002	40	1	41-50	Article
A test of genetic algorithms in relevance feedback	INFORM PROCESS MANAG	2002	38	6	793-805	Article
Automatic extraction of relationships between terms by means of Kohonen's algorithm	LIBR INFORM SCI RES	2002	24	3	235-250	Article
The effect of team consolidation on research collaboration and performance of scientists. Case study of Spanish university researchers in Geology	SCIENTOMETRICS	2002	55	3	377-394	Article
Spanish personal name variations in national and international biomedical databases: implications for information retrieval and bibliometric studies	J MED LIBR ASSOC	2002	90	4	411-430	Article
Vertical integration of sciences: an approach to a different view of knowledge organization	J INFORM SCI	2002	28	5	397-405	Article
Geographic information systems for science and technology indicators	RES EVALUAT	2002	11	3	141-148	Article
Assessing quality of domestic, scientific journals in geographically oriented disciplines: scientists' judgements versus citations	RES EVALUAT	2002	11	3	149-154	Article
Strategies and models for teaching how to use information: Guide for teachers, librarians and archivists	J ACAD LIBR	2002	28	1-2	77-78	Book Review

Fig. 6.8. List of the top 20 Spanish documents of the category Library Science and Information Science for the year 2002

6.5 Domain Visualization

We hold SVG to be a grand creation. It is a small, quick, ingenious format, and it is free. With firm technical support and broad acceptance on the part of the industry and experts, it was our clear choice as the format for visualization of the scientograms of vast scientific domains.

And we have no regrets. We must not forget that we wished to represent these major domains by means of JCR categories, which meant embedding at least 200 nodes, together with their corresponding links, into an A4-sized sheet of paper or the screen of a computer. And of course, the product had to be attractive and intelligible.

In order that the PFNET scientograms can be shown in vector format, we transform them to SVG format using ad hoc software in charge of making the final esthetic and informational touch-ups to the scientograms. This task includes the following:

SPAIN - COMPUTER SCIENCE, INFORMATION SYSTEMS - INFORMATION SCIENCE & LIBRARY SCIENCE

Title	Journal	Year	Vol.	N°	Pages	Doc. type
Automated code generation of dynamic specializations: an approach based on design patterns and formal techniques	DATA KNOWL ENG	2002	40	3	315-353	Article
Dual grid: A new approach for robust spatial algebra implementation	GEOINFORMATICA	2002	6	1	57-76	Article
The power of a pebble: Exploring and mapping directed graphs	INFORM COMPUT	2002	176	1	1-21	Article
Context-sensitive rewriting strategies	INFORM COMPUT	2002	178	1	294-343	Article
Practical algorithms for deciding path ordering constraint satisfaction	INFORM COMPUT	2002	178	2	422-440	Article
A fully syntactic AC-RPO	INFORM COMPUT	2002	178	2	515-533	Article
EIS evolution in large Spanish businesses	INFORM MANAGE-AMSTER	2002	40	1	41-50	Article
Parallel evolutionary algorithms can achieve super-linear performance	INFORM PROCESS LETT	2002	82	1	7-13	Article
Lower bounds on the information rate of secret sharing schemes with homogeneous access structure	INFORM PROCESS LETT	2002	83	6	345-351	Article
Document organization using Kohonen's algorithm	INFORM PROCESS MANAG	2002	38	1	79-89	Article
A test of genetic algorithms in relevance feedback	INFORM PROCESS MANAG	2002	38	6	793-805	Article
Bases for the development of LAST: a formal method for business software requirements specification	INFORM SOFTWARE TECH	2002	44	2	65-75	Article
Derived types and taxonomic constraints in conceptual modeling	INFORM SYST	2002	27	6	391-409	Article
Internet usage and competitive advantage: the impact of the Internet on an old economy industry in Spain	INTERNET RES	2002	12	5	391-401	Article
A context vector model for information retrieval	J AM SOC INF SCI TEC	2002	53	3	236-249	Article
Fuzzy ARTMAP and back-propagation neural networks based quantitative structure-property relationships (QSPRs) for octanol-water partition coefficient of organic compounds	J CHEM INF COMP SCI	2002	42	2	162-183	Article
Using molecular quantum similarity measures under stochastic transformation to describe physical properties of molecular systems	J CHEM INF COMP SCI	2002	42	2	317-325	Article
An integrated SOM-fuzzy ARTMAP neural system for the evaluation of toxicity	J CHEM INF COMP SCI	2002	42	2	343-359	Article
Modeling large macromolecular structures using promolecular densities	J CHEM INF COMP SCI	2002	42	4	847-852	Article
Polarizabilities of solvents from the chemical composition	J CHEM INF COMP SCI	2002	42	5	1154-1163	Article
Molecular Quantum Similarity-based QSARs for binding affinities of several steroid sets	J CHEM INF COMP SCI	2002	42	5	1185-1193	Article
In silico studies toward the discovery of new anti-HIV nucleoside compounds with the use of TOPS-MODE and 2D/3D connectivity indices. 1. Pyrimidyl derivatives	J CHEM INF COMP SCI	2002	42	5	1194-1203	Article
Parallel algorithms for graph cycle extraction using the cyclical conjunction operator	J CHEM INF COMP SCI	2002	42	6	1398-1406	Article

Fig. 6.9. Top 23 documents from Spain that cocite the categories Information Science & Library Sciences and Computer Sciences, Information Systems, year 2002

- Coloring the nodes with pre-determined colors.
- Coloring the links in red or blue, depending on whether their connection is "trustworthy" or not (see Sect. 6.4.1).
- Inserting a tag corresponding to the name of the category.
- Including the hyperlinks needed in the links and nodes to allow information associated with them to be retrieved.

The harvest consists of visualizations of large scientific domains that not only facilitate browsing around in the structure, but also make it easy to access and retrieve the bibliographic information around which they are constructed.

6.5.1 Evaluation

A series of requisites must be fulfilled in order to judge as well-designed the visualization of a domain (Börner; Chen; and Boyack, 2003).

Capacity to Represent Both Large and Small Amounts of Information

Thanks to the adoption of the *Kamada–Kawai* algorithm (1989), which distributes the information using the maximum space, reducing the number of crossed links and so on, we can offer scientograms that contain large as well as small quantities of information. We determined it possible to show in a clear fashion, on a regular sheet of paper or computer screen, up to 250 nodes and their tags. If no tags are included, as many as 500 nodes can be displayed. The adoption of the SVG format facilitates the implementation of certain complementary tasks of visualization, such as zooming in and amplifying or reducing any particular area of the scientogram to focus on zones of interest without losing a particle of quality from the original graphics; and at the same time, travelling within the graph in any direction.

Reduction of Time in the Visual Search for Information

The spatial distribution of the information, occupying the maximum space available, together with the manual repositioning of a tag or two if necessary, make the visual search of information very rapid – even in real time, with no need to resort to the zoom. Moreover, and again as a consequence of using SVG, we can quickly locate any chain of text we wish to, by means of a search tool incorporated in this format.

A Good Understanding of Complex Data Structures

The use of raw cocitation data values, in combination with the superposition of FA over the structure of a PFNET, makes the visualizations self-sufficient in relaying information (Fig. 6.10). Little interpretive effort is needed. The displays feature clusters of categories or groupings around one particular node, pointing to a high degree of relation among these disciplines, as well as an intensive intellectual interchange among them. In fact, special fields of study can be identified in this way. FA lets us perceive the larger grouping of categories or subject areas of science, and with different colors it highlights the positions, proximities, and categories of neighboring nodes. Finally, PFNET makes manifest the sematic structure of the domain, serving to thread together the different categories or disciplines that constitute it, and, by pruning, setting off the distinctive intellectual or subject groupings.

They Make Manifest Relationships of Which We Would Otherwise be Unaware

This objective is attained thanks to the use of social network theory, whose graphic representation allows us to show relations existing between pairs

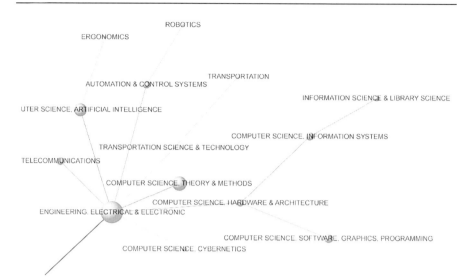

Fig. 6.10. Blow-up of an area of a fictitious PFNET domain

of nodes. Once again by using PFNET, we simplify the relations, showing only the strongest and most relevant ones, so that the structure of the domain is less complicated. The thickness and color of the links is indicative of the relation among the two categories involved, reflecting their degree of connection (thickness) and the reliability of the connection detected (color).

The Set of Data can be Observed From Different Perspectives

This is the only criterion with which we are not able to comply. SVG does not account for graphic rotations or panoramic viewing. Nonetheless, it may have these features in the future, as it is still under development and only one new version – 1.1 – has appeared since its creation.

It Favors the Formulation of Hypotheses

The visualizations proposed, which show the semantic/intellectual domain structure in an attractive and comprensible light, encourage even the non-expert to theorize about the area depicted, stirring up inferences about interactions that may come about in a certain context.

It Will Be the Object of Analysis, Debate, and Discussion

Our visualizations may be used by specialists as tools for analysis and debate about the current or past state of a domain. Hence, they may also be used, in sequential form, to study its evolution over time.

6.6 Scientography of Large Scientific Domains

The final result of the methodology put forth here is a scientogram capable of representing, on its own, the structure of great scientific domains with all the advantages described thus far, as revealed in Fig. 6.11.

In other words, we may say that the final product of our methodology is a set of scientograms that show the essential structure of science by means of the most significant links obtained from JCR category cocitation.

- They are user friendly, in terms of comprehension and handling, for a user from any background.
- They highlight the different scientific groupings that make up the structure of a domain, chaining them together in explicit sequences.
- They make manifest the semantic and intellectual structure of the domain they represent.
- They contribute with additional visual information, in nodes and links (an improvement over the PFNET created to date).
- They can be amplified, reduced, or shifted in any direction on the computer screen, without sacrificing quality.
- They can be easily displayed on the Web or sent by e-mail, in view of their low weight.

This methodology enables us to see, within a space as small as a simple sheet of paper, the structure of a vast domain, which in this case is geographical, but could be of any other type desired. This point is driven home by Fig. 6.11, which demonstrates all the above.

6.7 Intercitation

One of the possible criticisms this methodology might harvest is that it inherits the limitations of the classification system on which it is based – the JCR assignment of subject categories to its journals. Consequently, the scientograms may reflect the structure of the JCR itself, more than the science of a specific domain, which was our true intention. This accusation, if justified, would disqualify the use of category cocitation for depicting large scientific domains, and so it clearly calls for discussion.

In defiance of this claim, we wield two scientograms constructed using JCR category cocitation – one from journal cocitation and the other from document cocitation. The two will serve to demonstrate that our methodology

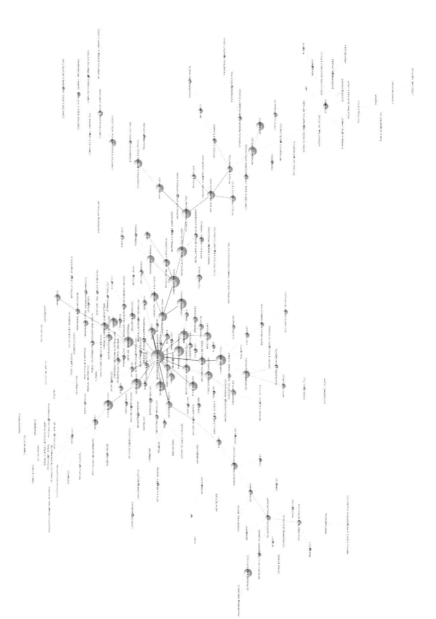

Fig. 6.11. PFNET scientogram of the structure of Spain's scientific domain in the year 2002

is fully valid, complying with the objectives for which it was proposed. Moreover, these scientograms arise as further tools for the study, comparison, and interpretation of science in general.

6.7.1 Journal Intercitation

Following the methodology proposed for the cocitation of categories, and using the subject area assignment of the JCR journals as the point of departure, we use PFNET to show the most significant relationships among categories and arrive at a scientogram that shows the essence of the purely scientific structure of the JCR (Fig. 6.12).

If we compare the scientograms of Figs. 6.11 and 6.12, our initial impression is that the two are structurally different. Yet a closer look brings to light certain similarities: both have *Biochemistry & Molecular Biology* as their central category; in both scientograms certain branches of the scientific structure share a pattern of connection, such as Neurosciences … Clinical Neurology … Psychiatry … Psychology; and many connections between categories coincide in the two scientograms, though not so many as to claim they are similar. Indeed, of all the possible connections represented in these two scientograms, those of Spain coincides with the global image of the JCR in 95 cases, and differs from it in another 119 (see Annex II).

The coincidence of links reveals a characteristic common to the two scientograms. If we transpose the concurrent categories onto the JCR structure we see that a great many are connected by thick links, indicating a strong degree of intercitation of the journals of those categories. It would follow that the reason behind the coincidence of links between categories is a high degree of intercitation of these journals, and that the scientific domain of a structure is nothing more than the reflection of the JCR itself, as a consequence of the strength of its relations with respect to the weakness of those established by the authors of the domain. But this is not true. If we look at the coincident links one by one, we notice that not all are strong links – there are also medium and weak ones, refuting the reasoning just given.

Yet it is certain that a series of coincident categories in the two scientograms exists, and there must be some explanation for this. The reason is, the links coincide because we are representing a single reality in different ways. That is, both the journal cocitation scientogram and the Spanish domain scientogram portray the same scientific structure, though more general in the case of intercitation; and the coincidence of the links merely evidences the structural similarities apparent from both perspectives. Therefore, we deduce that the two representations are structurally distinctive, although they share some specific points.

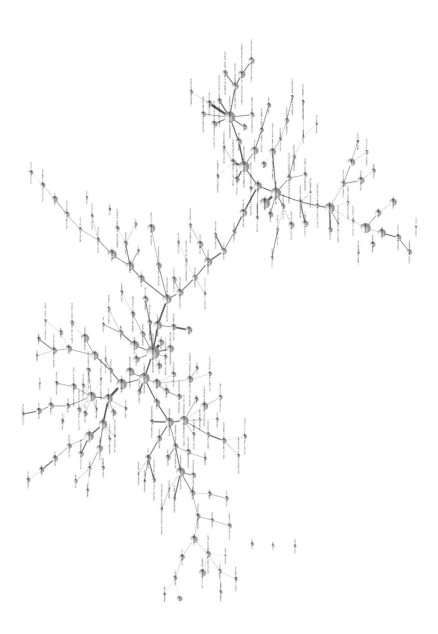

Fig. 6.12. Scientogram of the structure of the JCR 2002, based on its journals

The lack of coincidence of links has a very simple and logical explanation. The authors interact with science by means of their references to it. We take category cocitation and transform that interaction into graphic representations. When, however, author cocitations counteract those preestablished by journal intercitation, the relations between categories become visible only as the set of authors of a domain understand them, not as the JCR would predetermine them.

We have shown that the structure of a domain is not merely the reflection of the intercitation of JCR journals. Still, we must recall that the method used to construct our domain scientograms begins with document cocitation and goes onward to category cocitation. Thus, logic urges us to first corroborate if the structure of a domain depicted through category cocitation is as intended, and not just the spitting image of category intercitation extrapolated to documents.

6.7.2 Intercitation of Documents

To defend our methodology on all sides prone to attack, we must also compare the scientogram of a domain obtained through document intercitation with that same domain depicted by means of category cocitation as proposed here.

On the basis of the subject category assignment of the documents, according to JCR criteria (document →journal →JCR category), and in order to later show, through PFNET, the most significant relations among categories, we built a scientogram to depict the structure of JCR document intercitation (Fig. 6.13).

If we compare the scientogram of cocitation of the Spanish scientific domain – Fig. 6.11 – with the one for document intercitation, we quickly come to the conclusion that they also have very different structures. Or, more bluntly put, they are hardly even similar. This is remarkable, as the two are built with the same methodology and units of measure. The common characteristic previously seen, that of *Biochemistry & Molecular Biology* at the center of the scientograms (for the Spanish domain and for journal cocitation), does not take place here. It is likewise not easy to find chains of connections that are similar in these two scientograms. Yet the number of common links between the two is 112, whereas there are 102 different links (Annex II). And so we could say that they share connective aspects, while differing structurally.

As in the case of journal intercitation, if we translate the categories linked in both scientograms to the structure represented by document intercitation, some – but not all – were seen to be connected by strong links.

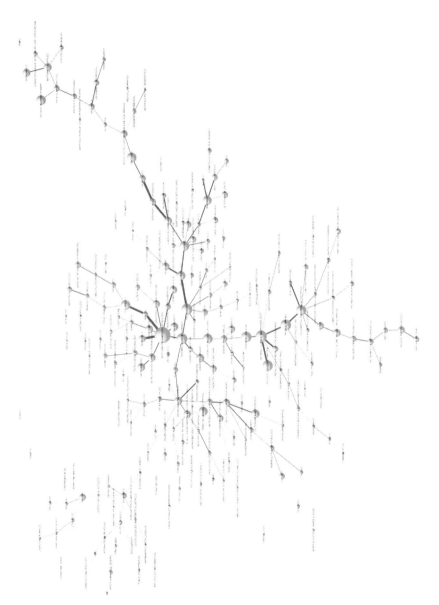

Fig. 6.13. Scientogram of the structure of the JCR, 2002, built on the basis of its documents

Again, this does not imply the prevalence of intercitation as opposed to cocitation, but rather, as we saw before, that one same object can be depicted from different viewpoints. The coincidence of links reveals some agreement despite the different perspectives.

6.7.3 Intercitation Versus Domain Scientograms

We stand at the crossroads of two different means of representing a single scientific structure: one depicted by JCR intercitation, the other by JCR cocitation. This gives rise to three types of scientograms, which we will call tangential ones: of journal intercitation, of document intercitation, and of category cocitation. Each offers a different yet complementary perspective of the same domain.

The scientogram of JCR journal intercitation is well consolidated on the scientific front, and corresponds with the mental scheme that society in general has. If we look at it carefully, we find no strange or weak links (in red) between categories, as occurs in the category cocitation for Spain. This is because it represents science from a retrospective and established perspective. In other words, it shows underlying relations after they have been accepted and adopted by the research community, the journals, and the JCR, and subsequently by the general public. And so, none of its links strikes us as being odd. For example, at one point in time the journals of LIS belonged only to the category Information Science & Library Science. Over time, however, as the publication of computer-related works increased, and the systems of information, bibliometrics, and so on became an everyday component of many fronts, researchers in the LIS field began to consider them inherent to their field – to the point where, when JCR classified some of the journals of this area as belonging to other categories as well, such as Computer Science and Information Systems, no one was surprised. Who knows whether they may someday belong to the areas of Computer Science, Software Graphics & Programming as well! To summarize, the JCR scientogram represents the consensus of the scientific community – including scientists themselves, editors, journal distributors, and the general readership.

The scientogram of document intercitation, while perhaps appearing a bit strange, offers a perspective of science from the viewpoint of the journals. It shows the most significant categories of the documents that they publish, evidencing multidisciplinary contents, and the most essential categories and relations. This scientogram can be situated halfway between that of journal intercitation and that of category cocitation: it is not as static and consolidated as the former, evolving as new contents are incorporated; but it is also not as dynamic as the latter. Its evolution does not only obey variations in the number of times that a category is assigned to the JCR journals, as is the case with journal intercitation, but depends as well on the number of documents published in a single journal of a specific category.

The scientogram of category cocitation of a given domain represents the structure of science from the author standpoint. It reflects the use that researchers make of scientific knowledge. It is therefore a changing and versatile scientogram in the relational sense, ideal for the study of the evolution of the scientific structure of a domain and for its comparison with others. It can cover long periods of time or short ones, even very short ones, depending on the interest at hand.

When comparing scientograms, their structural coincidences – in terms of links, patterns of connection and so forth – make manifest the most stable, permanent aspects of science. Here, we see that the more traditional model (intercitation) persists, alongside the more evolutionary cocitation model. Their differences signal the evolution of science in general and show which disciplines are changing and how they evolve, as reflected by the linked categories.

6.8 Analysis and Comparison of Domains

Our scientograms may be very useful in simplifying and displaying aspects otherwise hidden, yet they are admittedly complex. If not, they would not encourage analysis, one of our main objectives.

> The analysis of domains consists of making a map of the components or disciplines, fields or specialized areas most obvious within a general area of knowledge… where the maps are forms of automatic classification in which descriptors, documents, the complete works of an author, or journals, are represented (Hjørland, 2002).

In consonance with this statement by *Hjörland*, we built scientograms with the most obvious fields or specialties on the basis of their co-occurrences. In addition, we propose a methodology to study and compare, in detail, these scientograms.

Attempting to carry out a macrostructural analysis without any type of link pruning is practically impossible. The difficulty lies in the interdisciplinarity inherent to science, and the consequent multidisciplinary relations that emerge among the different categories. Virtually all the sciences are related in some way. Yet so many trees would not let us see the forest, as they say. In this case, an attempt to include the entire multitude of links would give Fig. 6.14.

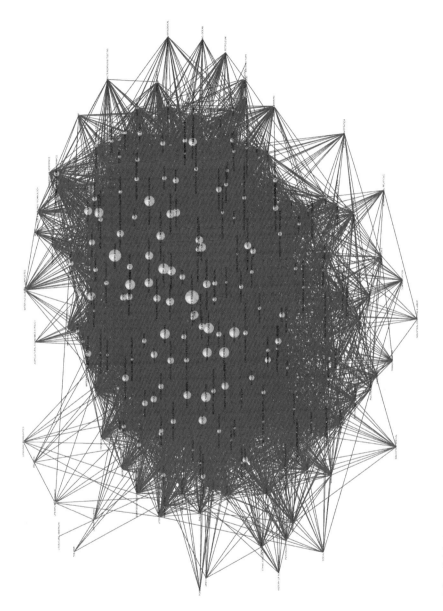

Fig. 6.14. Social network of the structure of Spain's scientific domain, 2002, with all its relations

Therefore, for improved visualization, comprehension and even better opportunities for browsing around the geographical macrostructure of the domain, we again resort to the PFNET scientograms, and not to the complete visualization of their structures with the totality of relations.

6.8.1 Intellectual Groupings

One of the doubts that may linger around the schematization of domains using PFNET is whether the display obtained faithfully reflects the structure and distribution of the original domain from which it comes. To confirm its correspondence, we carry out these steps:

- We perform FA of the original cocitation matrix.
- We indicate the number of factors detected with a variance greater than or equal to one.
- We show the elements of each factor and their accumulated variance.
- We denominate each factor according to the methods used by other authors.
- We revise and comment on the FA obtained.

Just below we transfer the above results to the PFNET scientogram, which calls for

- Assigning one same color to the categories – nodes – that make up each one of the different factors detected in the original matrix.
- Showing and identifying the number and percentage of factorized and non-factorized nodes.
- Detecting the possible dispersion of nodes of a single factor over the PFNET domain.

The execution of this process will allow us not only to show the degree of adequacy between the original matrix and the PFNET domain, but also to detect and identify the intellectual groupings in the form of major thematic areas, in addition to providing an image of the underlying semantic structure of the domain. This phenomenon can be seen in Fig. 6.15.

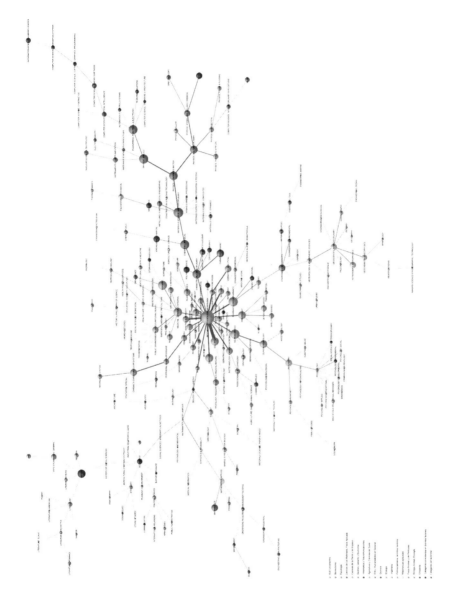

Fig. 6.15. Scientogram of the world Scientific Domain, 2002. Combination of PFNET and FA

6.8.2 Analysis of a Domain

Macrostructure

Having confirmed the correspondence between the original domain and its alter ego, we find ourselves able to secure a global view of the geographical structure of the domain, or its macrostructure. With this aim in mind, we take the following steps:

- We generate a new simplified scientogram by means of the graphic representation of the subject areas and their relations.
- We spatially situate each one of the subject areas, and establish their geographical relation with the factor scientogram they proceed from.
- We locate the basic macrostructural form through the groupings of the subject areas.
- We detect the central macrostructure, and comment on the reasoning behind this position and its consequences. We do the same for the groupings of peripheral thematic areas.
- We locate the central thematic area. We study its implications, repercussions on the domain, and effects on other subject areas.
- We indicate the geographical position of the rest of the subject areas and explain how their positioning influences the development of the domain.

Microstructure

The first task at hand is to describe in general terms the elements that make up the domain and its environs. To this end, we do the following:

- Indicate the total number of categories that make up the structure of each domain. We underline the fact that there are a few categories of great size – the highly productive ones – and many others with a small size or limited output. This reveals the hyperbolic nature that is typical of bibliometric distributions.
- We note whether all the categories are connected. If there are any isolated ones, we explain which, and the reason for their disconnection.
- We indicate if the model of connection of the categories is of the central–peripheral type. That is, if there is a central category that serves to connect others around it. We note whether there is a coincidence between the most central category and the most prominent one, by "prominent" meaning the one whose links make it particularly visible with respect to the others. In this case, we emphasize that this is a very common and necessary deed from the standpoint of social networks.

- By means of the measure of the degree of centrality, we detect the most central category. From there, and going over the paths of each category, we rank the proximity of each with respect to the center. Then, social network algorithms help us detect the most prominent nodes of the domain. We carefully consider the different measures obtained and comment on the results.

The Backbone

By means of cocitation, that is, through the observation of the chains produced in a scientogram between categories in terms of significant or thick links, we can spot the backbone of the domain, and study it separately if we wish. This is done by

- Identifying the thematic areas that make up the backbone.
- Detecting long-distance paths in each one of the subject areas.
- Taking note of the most relevant characteristics of the categories that make up this Backbone.

Surface Analysis

Exploring or retracing the geographical stretch over the surface of the domain, visiting the different nodes it comprises in terms of the connections or links that join it together allows us to shed light on the semantic structure of the domain, while also aiding with the explanation and description of the aspects related with the nodes and links we find en route. Thus,

- Before beginning the journey, we study the connections among categories, since they are the thread that strings together the main structural components. We detect the links of greater or lesser thickness, and use them to calculate the proportional intervals, on the basis of their degree of cocitation: strong, intermediate, or weak.
- Then, departing from the most central category and moving clockwise, beginning at high noon, we begin to explore the surface of the domain by visiting its links.
- As we go along the domain structure, we come up with explanations for the connections between categories. This is done by introducing queries in the links, so that we can access the information that gave rise to each intra-categorical connection, and justify the "strange bedfellows" – the red links – that we may encounter.
- Depending on how deep one wishes to delve into analysis, in order of relevance (Moya-Anegón et al., 2005), it is possible to determine to what extent each category contributes to the building of the link that

unites it with another node. This is tantamount to discovering the transfer of information between categories, or the contribution of one category to another: in technology, methodology, and so on. We can even find out to what extent the link that connects two categories is conceived indirectly by two categories other than the ones it unites.

The surface analysis, which is laborious, is not undertaken in this book. We merely divulge its methodology, which can be taken up in future excursions.

6.8.3 Domain Comparisons

Regardless of whether we intend to compare scientograms from different domains, or study the temporal evolution of some of them, we propose the following method for comparative studies.

Macrostructure

Just as we saw in the analysis of a single domain, it is important to confirm the equivalency between original structures and corresponding PFNET simplifications. Besides, in this way we manage to focus light on the intellectual structures and subject areas that constitute them. Once this correspondence is confirmed, we are able to compare macrostructures.

First, we compare the results obtained by applying FA to the original matrixes.

• We put, face to face, the factors detected in each domain and compare denomination, size, variance and order of each factor, and total of accumulated variance. This, from a numerical standpoint, takes us to an initial detailed approximation of the macrostructural differences of the scientograms under consideration.

After this, in order to obtain a graphic image with which to carry out visual comparisons, we transfer the results of PCA to the scientograms at hand. In doing this, we assign one same color to the nodes that make up each one of the different factors detected in the original matrix.

We indicate the number of factorized and non-factorized categories, giving them a different color to distinguish them from the rest.

• We contrast to see if all the domains or evolutive scientograms have the central–peripheral model of factor connection, or try to identify some alternative pattern of connection.
• We establish geographical coincidences between the factors of the different scientograms. We determine whether these positions are relative or

absolute, and if possible, which are always relative and which are consistently absolute. Specialized areas will be detected by the identification of smaller factors within larger factors. Moreover, in the evolutive scientograms, we locate factors whose nodes are scattered throughout the scientogram, so as to observe their evolution.

- Through the measures of centrality, we identify the most central factor; and using specific algorithms, we identify the most prominent factors of each domain and their relevance. We compare the different measures and look for signs of evolution.
- We indicate and study the paths of connection between factors to detect whether there are alterations in the sequences of categories.
- We identify the points of confluence of factors. This means detecting those nodes with an ascription to multiple factors, as these are the bridges or points of interchange of knowledge among factors. We establish whether they are the same in the different scientograms and whether they "put in touch" the same factors in the different domain scientograms.
- We seek out changes of node assignment in the factors of the different domains and observe their evolution. This stage is very important for studying disciplinary changes over time in one or more domains.

Microstructure

The microstructural comparative analysis of several domains, or of one over time, entails contrasting the general elements of each domain and of their surroundings.

- To determine the evolution or the changes in its output per category, we compare the number of nodes that make up each domain. We put special emphasis on the appearance of new nodes or the disappearance of former nodes. If desired, we can compare the output of specific categories.
- We detect disconnections in the domains, and elucidate the causes. To do so, we identify the isolated categories of each scientogram, indicate if they are the same in the different domains or evolutive scientograms, and explain the reasons for their isolation.
- We corroborate if, in all the domains or evolutive scientograms, the model of category interconnection is of the central–peripheral type.
- Using the measures of centrality, we identify the most central categories of each domain and the distance of the rest with respect to these central points. Specific algorithms then allow us to identify the most prominent categories of each domain. We compare the different means and study their evolution, if any.
- We look into the nature of the prominent categories that constitute the source of a thematic area.

The Backbone

Having identified and extracted the most relevant sequences of each one of the domains for comparative analysis, we proceed to contrast

- the thematic areas that comprise the Backbone or central nervous system of each domain;
- the longest paths of the given domain;
- and the most significant categories of each domain, as well as their main characteristics.

Surface Comparison

Again, the surface terrain means a rugged journey if we wish to explore, on a microlevel, the links of the different scientograms. A less laborious shortcut would allow us to at least compare the overall microstructures of the different domains. As before, we merely expose the methodology here, so that it might be undertaken in full detail through further studies.

To compare domain surfaces, we

- detect connections between categories that remain invariable in the different domains or evolutive scientograms under consideration;
- detect and identify categories whose links change from one scientogram to the next, and study their evolution.

If so desired, for example in order to explain a very strange or surprising connection between categories, or variations in their linking in different scientograms, we may resort to a detailed analysis of the connections among nodes, as well as the study of the direction of transfer of information between them.

6.8.4 Evolution of a Domain

General consensus has it that both in the world of information visualization and that of geographical cartography, several maps of one same object of study are always preferable to a single map. This is so because each map provides a different perspective.

The precursor of the study of science by means of series of maps was *Garfield* (1994), with his introduction of the concept of longitudinal maps. These consist of a series of chronological and sequential maps, from which it is possible to deduce the evolution of scientific knowledge.

The process that we shall adopt for the evolutive study does not differ substantially from that presented in Section 6.8.3, and so we do not reproduce it here.

7 Results

The results are presented in three main sections. The first is the scientography and analysis of a major geographic domain: the world. The second
holds the comparative analysis of two great geographic domains: the
United States and the European Union. And the third is the evolutive study
of another large domain, Spain.

7.1 Scientography and Analysis of the World, 2002

7.1.1 Scientogram

The scientogram in Fig. 7.1 shows world scientific output on the basis of
ISI categories. This scientogram is the visualization obtained by applying
our methodology, and it graphically represents the 901,493 scientific documents gathered in the ISI databases, grouped into the 218 JCR categories
showing production in the year 2002. The correspondence between each
one of these categories and its output can be seen in Table 7.1.

In order to favor comprehension of the scientogram, each sphere has
been tagged with the name of the JCR category it represents, and is given a
size that varies, being directly proportional to the amount of documents it
holds. To help establish the visual association between the size of each
category and its true production, the lower left part of the scientogram
gives a sphere of reference, with a size equivalent to 1,000 documents. The
lines that connect the different spheres are the relations of cocitation
between categories that are most significant or essential, as the superfluous
ones have been eliminated by PFNET. These relations also vary in thickness (greater intensity means greater thickness), and they come to represent the consensual point of view of 1,751,996 authors worldwide, by
means of their 25,682,754 citations.

The spatial distribution of the categories in the scientogram was
achieved using the Kamada–Kawai algorithm. The structural disposition is
determined by the tandem: raw category cocitation, in combination with
PFNET.

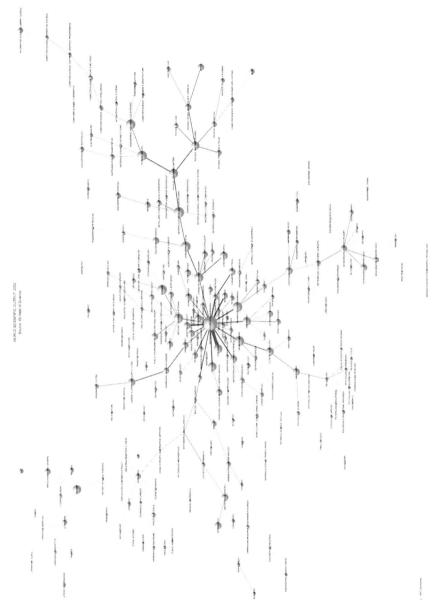

WORLD SCIENTIFIC OUTPUT 2002
Source: ISI Web of Science

Fig. 7.1. World scientogram, 2002

Table 7.1. Listing of JCR categories and world output, 2002

Category	Prod.	Category	Prod.
Energy & Fuels	4826	Medicine, Legal	1004
Geosciences, Interdisciplinary	11812	Anthropology	3100
Engineering, Petroleum	1527	Peripheral Vascular Disease	9439
Gastroenterology & Hepatology	9356	Rehabilitation	2927
Radiology, Nuclear Medicine & Medical Imaging	11441	Sport Sciences	4933
Mathematics	13668	Instruments & Instrumentation	7366
Medicine, General & Internal	22619	Geochemistry & Geophysics	8730
Education, Scientific Disciplines	1451	Mineralogy	1715
Medical Informatics	1212	Engineering, Environmental	4278
Entomology	3963	Computer Science, Hardware & Architecture	2951
Chemistry, Multidisciplinary	22598	Developmental Biology	3879
Construction & Building Technology	1948	Andrology	344
Materials Science, Multidisciplinary	28804	Engineering, Biomedical	4384
Computer Science, Theory & Methods	7146	Oceanography	5731
Computer Science, Information Systems	4692	Psychology	3635
Computer Science, Software, Graphics, Programming	4504	Nuclear Science & Technology	7150
Mathematics, Applied	13509	Chemistry, Applied	5868
Acoustics	3326	Physics, Fluids & Plasmas	5918
Cardiac & Cardiovascular Systems	14392	Information Science & Library Science	8096
Respiratory System	6753	Physics, Particles & Fields	8385
Agriculture, Dairy & Animal Science	4441	Chemistry, Medicinal	6148
Agriculture	4799	Materials Science, Paper & Wood	1039
Agriculture, Soil Science	2695	Ergonomics	584
Nutrition & Dietetics	5775	Spectroscopy	6049
Food Science & Technology	8184	Operations Research & Management Science	3774
Anesthesiology	5229	Engineering, Marine	149
Anatomy & Morphology	1327	Thermodynamics	3452
Engineering, Aerospace	2510	Materials Science, Coatings & Films	3210
Astronomy & Astrophysics	12279	Fisheries	3300
Biochemistry & Molecular Biology	54972	Limnology	1240

Table 7.1. (Cont.)

Category	Prod.	Category	Prod.
Plant Sciences	13869	Mathematics, Miscellaneous	1009
Zoology	7116	Geography	1867
Biology	5763	Ornithology	988
Biotechnology & Applied Microbiology	13477	Materials Science, Biomaterials	1555
Surgery	24490	Geology	1605
Crystallography	6277	Computer Science, Interdisciplinary Applications	5709
Biochemical Research Methods	8749	Horticulture	2209
Cell Biology	19834	Computer Science, Cybernetics	866
Dermatology & Venereal Diseases	5902	Transplantation	4546
Endocrinology & Metabolism	12054	Psychology, Experimental	3367
Hematology	10977	Psychology, Mathematical	421
Environmental Sciences	14394	Materials Science, Characterization & Testing	1227
Water Resources	5729	Engineering, Civil	4991
Marine & Freshwater Biology	6881	Mining & Mineral Processing	1474
Metallurgy & Metallurgical Engineering	8422	Physics, Mathematical	7856
Mechanics	9436	Electrochemistry	3633
Medicine, Research & Experimental	9694	Psychology, Developmental	2670
Neurosciences	26350	Education & Educational Research	4337
Clinical Neurology	18324	Engineering, Industrial	2796
Pathology	7041	Mycology	1237
Obstetrics & Gynecology	7857	Psychology, Educational	1310
Dentistry, Oral Surgery & Medicine	4990	Environmental Studies	2913
Ecology	7837	Social Sciences, Biomedical	898
Oncology	20330	Psychology, Biological	1144
Ophthalmology	6396	Sociology	4774
Orthopedics	5528	Law	3518
Otorhinolaryngology	3705	Microscopy	871
Pediatrics	10465	Remote sensing	1071
Pharmacology & Pharmacy	25888	Transportation	332
Physics, Multidisciplinary	18056	Psychology, Clinical	4384
Physiology	9919	Imaging Science & Photographic Technology	958
Polymer Science	11201	Materials Science, Textiles	958
Physics, Applied	25579	Economics	9053
Engineering, Chemical	12407	Social Sciences, Mathematical Methods	1247
Psychiatry	13953	Business	2822

Table 7.1. (Cont.)

Category	Prod.	Category	Prod.
Tropical medicine	1517	Management	3253
Parasitology	2384	Urban Studies	1422
Veterinary Sciences	10662	Social Sciences, Interdisciplinary	2423
Virology	4649	Planning & Development	2046
Substance Abuse	1801	Business, Finance	1594
Engineering, Mechanical	8464	Social Issues	1104
Microbiology	13210	Public Administration	1191
Statistics & Probability	4609	Social Work	1354
Physics, Nuclear	6102	Psychology, Social	2206
Optics	11449	Nursing	2459
Physics, Atomic, Molecular & Chemical	11526	Women's Studies	1138
Biophysics	10507	Area Studies	2830
Chemistry, Organic	17006	Political Science	5526
Chemistry, Analytical	13445	Criminology & Penology	1023
Medical Laboratory Technology	2455	International Relations	2569
Chemistry, Physical	24789	Education, Special	670
Materials Science, Composites	2789	Archeology	1560
Reproductive Biology	3951	History	20147
Genetics & Heredity	13023	Family Studies	1498
Immunology	18431	History Of Social Sciences	1357
Engineering, Electrical & Electronic	24042	Psychology, Psychoanalysis	870
Chemistry, Inorganic & Nuclear	10402	Language & Linguistics	3045
Physics, Condensed Matter	21990	Psychology, Applied	1968
Automation & Control Systems	3365	Health Policy & Services	2397
Meteorology & Atmospheric Sciences	7581	Philosophy	4319
Behavioral Sciences	3967	Ethnic Studies	368
Toxicology	7236	Industrial Relations & Labor	804
Geriatrics & Gerontology	2502	Communication	1578
Engineering	4199	Demography	768
Agricultural Economics & Policy	411	Engineering, Geological	1214
Forestry	2406	Health Care Sciences & Services	4094
History & Philosophy Of Science	2164	Literature, Romance	3864
Computer Science, Artificial Intelligence	6805	Theater	1423
Engineering, Manufacturing	3329	Literary Reviews	3956
Infectious Diseases	8095	Literature, Slavic	373
Rheumatology	3394	Engineering, Ocean	752
Urology & Nephrology	8818	Music	7227
Telecommunications	4275	Literature	7048
Paleontology	1566	Poetry	558
Allergy	2203	Literature, American	657
Biology, Miscellaneous	2276	Arts & Humanities, General	10170

Table 7.1. (Cont.)

Category	Prod.	Category	Prod.
Materials Science, Ceramics	4675	Architecture	1493
Public, Environmental & Occupational Health	12321	Art	6189
Emergency Medicine & Critical Care	1403	Religion	5769

   The raw cocitation plus PFNET tandem makes the thematic areas crop out around prominent categories as if they were clusters of grapes, and also highlights the chains of explicit sequences they form. The order of the categories that produce this chain is not arbitrary, but reveals how the bunches or subject areas are interconnected. In this way, the connections among categories and the prominent categories delimit thematic areas, and the connections between prominent categories reveal how the subject areas are united. For instance, with a quick look at Fig. 7.1, we can distinguish a large central bunch surrounded by smaller clusters, distributed all over the surface of the scientogram. If we look more closely at the central bunch and at one of those in the lower section, we discover the chain Biochemistry & Molecular Biology ←→ Neurosciences ←→ Clinical Neurology ←→ Psychiatry ←→ Psychology. This path indicates that in the scientogram there appear two major thematic areas we could call Biomedicine and Psychology, whose most prominent categories are Biochemistry & Molecular Biology, and Psychology, respectively, which are connected by other intermediate categories such as Neurosciences, Clinical Neurology, and Psychiatry. We could say the same, to give just one more example, of the sequence at the left central section: Mathematics Miscellaneous ←→ Social Sciences Mathematical Methods ←→ Economics ←→ History of Social Sciences ←→ History, which shows how two larger areas we will refer to as Mathematics and Humanities are united. These paths are very important, as they bear the thread that strings together the entire scientific structure of a domain.
   Clearly this tandem captures the essence of the domain structure while also revealing the subject areas that make it up. Yet it does not identify or delimit them, as other statistical techniques such as FA would do.

7.1.2 Factor Analysis

The detection of the principal thematic areas of the scientogram is achieved through FA. We apply a variant of FA known as PCA, with a varimax rotation – where the missing values are replaced by means – upon the original cocitation matrix of 218 x 218 categories, with raw cocitation values.

The FA identifies 33 factors in the cocitation matrix of 218 x 218 categories. The criteria adopted to stop the extraction of factors was the appearance of an eigenvalue greater than or equal to one.[3] Of the 33 factors identified, we extracted 16, in agreement with the scree test,[4] which accumulated 70.2% of the variance (Table 7.2).

In order to capture the nature of each factor and categorize it, we followed the methodology proposed by *Moya-Anegón, Jiménez Contreras* and *Moneda Carrochano* (1998). First, the factors were ordered by their weighting index or factor loading in decreasing order. We established as the limit for belonging a value equal to or greater than 0.5, whereas for denomination we only took into account the categories of each factor with a value of at least 0.7.

Table 7.2. Eigenvalues of the first 16 factors corresponding to the world, domain, 2002

Factor	Eigenvalue	% variance	% of cumulative variance
1	42.255	19.4	19.4
2	24.14	11.1	30.5
3	15.472	7.1	37.6
4	12.655	5.8	43.4
5	10.069	4.6	48
6	8.272	3.8	51.8
7	6.815	3.1	54.9
8	6.298	2.9	57.8
9	4.668	2.1	59.9
10	4.517	2.1	62
11	4.195	1.9	63.9
12	3.601	1.7	65.6
13	3.029	1.4	67
14	2.567	1.2	68.1
15	2.321	1.1	69.2
16	2.16	1	70.2

[3] This criterion works quite well, offering results much in agreement with the expectations of researchers (Ding et al., 1999)
[4] The *scree test* consists of examining the line obtained with the graphic representation of the Eigenvalues of identified factors. The extractions of factors is stopped when the eigenvalue line begins to level out, practically parallel to the axis (Lewis-Beck, 1994)

Factor 1: Biomedicine

Factor 1 is the greatest of the 16 factors identified as far as the number of categories is concerned. In itself, it explains 19.4% of variance. It is made up of 63 categories that represent 28.9% of the 218 categories making up the cocitation matrix. Attempting to categorize this factor in view of the 39 categories that we considered useful for denomination, with a value of at least 0.7 (the italicized ones alone), was no easy task in the face of such diversity. But careful observation of the categories, and the inference of their equivalents as scientific disciplines, whenever possible, led us to the conclusion that the best denomination for this factor was *Biomedicine* (Table 7.3)

Table 7.3. Factors 1 and 2 with factor loadings greater than or equal to 0.5

Factor 1	% variance	Factor 2	% variance
Biomedicine	19.4	Psychology	11.1
Endocrinology & Metabolism	0.951	*Psychology, Social*	0.915
Medicine, Research & Experimental	0.948	*Psychology, Developmental*	0.851
Oncology	0.936	*Family Studies*	0.832
Pathology	0.929	*Social Work*	0.823
Urology & Nephrology	0.923	*Criminology & Penology*	0.812
Medical Laboratory Technology	0.914	*Psychology, Clinical*	0.810
Physiology	0.907	*Psychology, Educational*	0.806
Immunology	0.891	*Psychology, Applied*	0.805
Gastroenterology & Hepatology	0.890	*Education & Educational Research*	0.798
Dermatology & Venereal Diseases	0.883	*Psychology, Psychoanalysis*	0.795
Pharmacology & Pharmacy	0.878	*Women's Studies*	0.780
Biophysics	0.878	**Substance Abuse**	0.692
Biology	0.874	**Psychology, Experimental**	0.667
Developmental Biology	0.861	**Social Sciences, Interdisciplinary**	0.656
Genetics & Heredity	0.860	**Psychology**	0.650
Nutrition & Dietetics	0.858	**Communication**	0.594
Ophthalmology	0.855	**Social Sigues**	0.574
Geriatrics & Gerontology	0.851	**Ergonomics**	0.539
Rheumatology	0.850	**Psychology, Mathematical**	0.539
Anatomy & Morphology	0.848	**Psychiatry**	0.538
Cell Biology	0.834	**Social Sciences, Biomedical**	0.537
Virology	0.823	**Education, Special**	0.536
Hematology	0.821		

Table 7.3. (Cont.)

Factor 1	% variance	Factor 2	% variance
Medicine, General & Internal	0.812		
Pediatrics	0.810		
Biotechnology & Applied Microbiology	0.806		
Toxicology	0.804		
Neurosciences	0.793		
Andrology	0.782		
Biochemical Research Methods	0.759		
Veterinary Sciences	0.753		
Microbiology	0.752		
Reproductive Biology	0.751		
Respiratory System	0.749		
Microscopy	0.748		
Obstetrics & Gynecology	0.746		
Biochemistry & Molecular Biology	0.724		
Dentistry, Oral Surgery & Medicine	0.721		
Surgery	0.711		
Peripheral Vascular Disease	0.685		
Radiology, Nuclear Medicine & Medical Imaging	0.685		
Plant Sciences	0.669		
Cardiac & Cardiovascular Systems	0.663		
Parasitology	0.659		
Chemistry, Analytical	0.642		
Transplantation	0.636		
Mycology	0.620		
Public, Environmental & Occupational Health	0.612		
Allergy	0.608		
Zoology	0.605		
Otorhinolaryngology	0.599		
Sport Sciences	0.597		
Clinical Neurology	0.594		
Infectious Diseases	0.591		
Chemistry, Medicinal	0.583		
Medicine, Legal	0.575		
Anesthesiology	0.558		

Table 7.3. (Cont.)

Factor 1	% variance	Factor 2	% variance
Biology, Miscellaneous	0.544		
Engineering, Biomedical	0.527		
Entomology	0.511		
Emergency Medicine & Critical Care	0.510		
Agriculture, Dairy & Animal Science	0.506		
Food Science & Technology	0.501		

Factor 2: Psychology

Comprising 22 categories, 10.1% of the total, this factor stands for 11.1% of the variance. In light of the first 11 categories (those with a factor loading of 0.7 or more) the denomination of this factor is obvious.

Factor 3: Materials Sciences and Applied Physics

Factor 3 is made up of 22 categories. It accumulates 7.1% of total variance, and 10.1% of the categories that constitute the matrix. Looking at the first 11, those with a value of 0.7 or more, we see that all are very much related with material sciences and applied physics, for which reason we decided to categorize this factor as the union of these two concepts: *Materials Science and Applied Physics* (Table 7.4).

Table 7.4. Factors 3 and 4, with factor loadings greater than or equal to 0.5

Factor 3	% variance	Factor 4	% variance
Materials Sciences and Applied Physics	7.1	Earth & Spaces Sciences	5.8
Materials Science, Ceramics	0.932	*Geology*	0.912
Materials Science, Coatings & Films	0.873	*Paleontology*	0.895
Metallurgy & Metallurgical Engineering	0.859	*Meteorology & Atmospheric Sciences*	0.879
Physics, Applied	0.822	*Oceanography*	0.878
Physics, Condensed Matter	0.818	*Astronomy & Astrophysics*	0.875
Polymer Science	0.781	*Geochemistry & Geophysics*	0.844
Chemistry, Physical	0.771	*Mineralogy*	0.740
Materials Science, Multidisciplinary	0.757	*Engineering, Geological*	0.736

Table 7.4. (Cont.)

Factor 3	% variance	Factor 4	% variance
Electrochemistry	0.742	*Engineering, Ocean*	0.735
Physics, Atomic, Molecular & Chemical	0.729	*Remote Sensing*	0.729
Physics, Multidisciplinary	0.717	*Geosciences, Interdisciplinary*	0.722
Instruments & Instrumentation	0.684	**Engineering, Aerospace**	0.673
Optics	0.676	**Geography**	0.635
Materials Science, Characterization & Testing	0.670	**Limnology**	0.567
Crystallography	0.668	**Engineering, Petroleum**	0.546
Mining & Mineral Processing	0.658		
Engineering, Electrical & Electronic	0.627		
Spectroscopy	0.576		
Mechanics	0.566		
Nuclear Science & Technology	0.551		
Chemistry, Multidisciplinary	0.541		
Engineering, Chemical	0.518		

Factor 4: Earth and Space Sciences

This factor evokes 5.8% of the variance. It holds 15 categories, or 6.9% of the total, which are distributed thus: 11 have values of 0.7 or more, and just 4 reside below that threshold. Again, looking at the categories that are italicized, the denomination of this factor is simple.

Factor 5: Management, Law and Economics

With its share of 4.6% of the total accumulated variance, and 20 categories (9.17% of the total), 9 of which might not be adequate for its categorization (in bold), this factor was not so easy to name, in view of its disciplinary diversity and its being a conglomerate of categories pertaining to or closely related to the social sciences; and so the choice and combination of terms that allowed us to embrace the whole thematic area of this factor took some time. After much consideration, we believe this denomination to be the most appropriate (Table 7.5).

Factor 6: Computer Science and Telecommunications

Possibly the easiest factor to categorize. Its eight categories with a value of 0.7 or over indicate this much. We might have even added the term "automatization" to the factor name, but in the end we decided that this notion was implicit in Computer Science. This factor accumulates 3.8% of the total variance, and its nine categories comprise 4.13% of the total categories in the cocitation matrix. Noteworthy is, precisely, the last of these nine categories – Information Science & Library Science – which does not have such a high index so as to lend it for use in the denomination of the factor, but does qualify it for membership. Also remarkable is the fact that, in Spain at least, debate has long revolved around the pertinence of LIS to the Social Sciences or to Arts and Humanities. The methodology used here

Table 7.5. Factors 5 and 6 with factor loadings greater than or equal to 0.5

Factor 5	% variance	Factor 6	% variance
Management, Law and Economics	4.6	Computer Science and Telecommunications	3.8
Planning & Development	0.877	*Computer Science, Artificial Intelligence*	0.905
Industrial Relations & Labor	0.861	*Computer Science, Hardware & Architecture*	0.904
Public Administration	0.855	*Computer Science, Information Systems*	0.849
Area Studies	0.824	*Computer Science, Theory & Methods*	0.839
Political Science	0.787	*Computer Science, Software, Graphics, Programming*	0.834
International Relations	0.771	*Telecommunications*	0.791
Agricultural Economics & Policy	0.763	*Automation & Control Systems*	0.781
Law	0.744	*Computer Science, Cybernetics*	0.727
Urban Studies	0.736	**Information Science & Library Science**	0.551
Business, Finance	0.697		
History Of Social Sciences	0.685		
Sociology	0.670		
Environmental Studies	0.632		
Ethnic Studies	0.629		
Social Sigues	0.611		
Demography	0.540		
History	0.529		
Communication	0.518		
Social Sciences, Interdisciplinary	0.516		
Economics	0.501		

settles the matter, as does the factor scientogram we shall see a bit later, in that the thematic area where it is best integrated is *Computer Science*.

Factor 7: Agriculture and Soil Sciences

With its 3.1% of the total variance and its four categories, just 1.83% of the total, two of which reach the value of 0.7, this factor was difficult to categorize. While at first we considered terms such as Horticulture and Agriculture (the two categories in italics) we finally opted for this title, which represents a mean value slightly below 0.7 (Table 7.6).

Factor 8: Humanities

Consisting of 12 categories, 5.5% of the whole, it explains 2.9% of the variance. Given that the first six categories, all with a loading factor of 0.7 or more, have to do with Literature, Linguistics, Art and Poetry, we justify this generic denomination as the most adequate.

Table 7.6. Factors 7, 8, and 9, with factor loadings greater than or equal to 0.5

Factor 7	% variance	Factor 8	% variance	Factor 9	% variance
Agriculture and Soil Sciences	3.1	Humanities	2.9	Chemistry	2.1
Horticulture	0.823	*Literary Reviews*	0.882	*Chemistry, Organic*	0.748
Agriculture	0.763	*Literature, American*	0.808	*Chemistry, Applied*	0.701
Forestry	0.699	*Language & Linguistics*	0.806	**Chemistry, Inorganic & Nuclear**	0.679
Agriculture, Soil Science	0.696	*Theater*	0.737	**Engineering, Chemical**	0.622
		Arts & Humanities, General	0.736	**Chemistry, Medicinal**	0.599
		Poetry	0.727	**Materials Science, Textiles**	0.597
		Literature, Romance	0.683	**Crystallography**	0.520
		Literature	0.661	**Polymer Science**	0.520
		Art	0.647		
		Religion	0.629		
		Philosophy	0.553		
		History	0.548		

Factor 9: Chemistry

The ninth factor is made up of eight categories. It accounts for 2.1% of the total variance and 3.67% of the categories in the cocitation matrix. Its first two categories, and even the third, have a value very close to 0.7, so we chose to categorize this factor in a rather general way, to take in as many categories as possible.

Factor 10: Etiology

Accounting for 2.1% of variance, this factor comprises three categories, or 1.37% of the total. Whereas the category with a value over 0.7 is Behavioral Sciences, and this was considered for denomination, the two accompanying categories – Psychology, Biological and Psychology Experimental – made us reflect on the reference to the study of animal conduct more than human behavior. Further consideration proved this to be so. The journals assigned to these categories publish work related with the study of animal behavior in their natural habitat, in other words, etiology (Table 7.7).

Factor 11: Mechanical Engineering

Integrating eight categories, 3.67% of the whole, it gathers up 1.9% of the variance. This factor has just two categories over the 0.7 threshold, yet in view of their contents, we opted for the title of the component with the higher value. Its categories are very near the 0.7 cutoff; and some might

Table 7.7. Factors 10 and 11, with factor loadings greater than or equal to 0.5

Factor 10	% variance	Factor 11	% variance
Etiology	2.1	Mechanical Engineering	1.9
Behavioral Sciences	0.701	*Engineering, Mechanical*	0.771
Psychology, Biological	0.633	*Engineering*	0.728
Psychology, Experimental	0.549	**Thermodynamics**	0.681
		Materials Science, Composites	0.668
		Materials Science, Characterization & Testing	0.619
		Acoustics	0.578
		Construction & Building Technology	0.561
		Mechanics	0.518

have helped categorize this factor, yet we chose to overlook them, as they already participated in another factor, with a high capacity of categorization (specifically, Factor 1 3).

Factor 12: Health Policy, Medical Services

With 1.7% of the total variance accumulated and six categories – 2.75% of all – this factor has three categories that suffice for its categorization (Table 7.8).

Factor 13: Applied Mathematics

Despite having seven categories, representing 3.21% of the whole, this factor only explains 1.4% of variance. Furthermore, it features just two categories in italics, from which to categorize the rest. We therefore chose to let in some influence from the rest of the categories of the factor. The common denominator in the majority is the term "mathematics"; and where it does not appear per se, we find other terms such as "economics" or "finance". This led us to the denomination Applied Mathematics.

Factor 14: Nuclear Physics and Particle Physics

With six categories, it sums up just 1.2% of the variance and represents 2.75% of all the categories. The disparity insofar as the contents of one of the three categories with a factor loading of 0.7 or more made it difficult to categorize this factor, but we believe we settled on the most adequate denomination (Table 7.9).

Table 7.8. Factors 12 and 13 with factor loadings greater than or equal to 0.5

Factor 12	% variance	Factor 13	% variance
Health Policy, Medical Services	1.7	Applied Mathematics	1.4
Health Policy & Services	0.808	*Social Sciences, Mathematical Methods*	0.745
Nursing	0.773	*Statistics & Probability*	0.725
Social Sciences, Biomedical	0.737	**Mathematics, Miscellaneous**	0.664
Health Care Sciences & Services	0.680	**Psychology, Mathematical**	0.630
Medical Informatics	0.622	**Economics**	0.580
Tropical Medicine	0.577	**Business, Finance**	0.579
		Mathematics, Applied	0.515

Table 7.9. Factors 14, 15, and 16 with factor loadings greater than or equal to 0.5

Factor 14	% variance	Factor 15	% variance	Factor 16	% variance
Nuclear Physics and Particle Physics	1.2	Animal Biology, Ecology	1.1	Orthopedics	1
Physics, Mathematical	0.753	*Ornithology*	0.781	*Orthopedics*	0.748
Physics, Nuclear	0.745	**Entomology**	0.646	**Otorhinolaryngology**	0.617
Physics, Particles & Fields	0.743	**Ecology**	0.622	**Emergency Medicine & Critical Care**	0.586
Physics, Fluids & Plasmas	0.674	**Anthropology**	0.602		
Optics	0.521	**Zoology**	0.575		
Education, Scientific Disciplines	0.504	**Fisheries**	0.563		
		Marine & Freshwater Biology	0.552		
		Biology, Miscellaneous	0.506		

Factor 15: Animal Biology, Ecology

Made up of six categories that constitute 2.75 of the total and only 1.1% of the variance, this factor denomination is based on just one category. However, in view of the categories that follow, with factor loadings nearly reaching 0.7, we believed it best to broaden this subject area so that it would not be limited to only one special field.

Factor 16: Orthopedics.

It is the last of the factors we extracted. Comprising three categories that stand for 1.37 % of the total, it explains 1% of variance. It has only one category with a value equal to or greater than 0.7, so its denomination was automatic. In this case, however, Orthopedics should be understood from the Anglo-Saxon perspective, as orthopedic surgery: specialized medicine related with the preservation and restoration of the skeletal system and its associated structures, such as the Backbone and other bones, joints, and muscles (Encyclopaedia Britannica, 2005). This meaning justifies the inclusion of the following category, Otorhinolaryngology, closely related with dental implants and prosthesis (osteo-system) and the same can be said of Emergency Medicine and Critical Care, concerned with the preservation of bones, muscles, and joints. This clarification is made because the same factor is reproduced in other domains, but with a different ordering of categories, where it will receive the same denomination.

Factor Scientogram

The transfer of factors identified by means of FA to the scientogram structure facilitated by PFNET allows us to determine the level of coincidence among the thematic areas detected by one method or the other. But that is not all. It also favors the visualization of these subject areas, providing an image of the underlying intellectual structure of the domain through what we call factor scientogramming.

Out of the 218 categories that make up the scientogram of the world, 195 are identified by FA. At a glance we can locate each of the thematic areas and the categories that they contain, colored in the same tone. Those that are not identified and do not belong to any particular thematic area are shown is dark gray.

In the lower left zone of Fig. 7.2 there is a figure legend establishing the color of each thematic area and the name given to each (explained in a previous section). To aid visibility and an understanding of the relation between the size of each category and its true production, a sphere of reference

Fig. 7.2. Factor scientogram of the world, the year 2002

equivalent to 1,000 documents is included. Finally, the categories belonging to more than one thematic area are colored in red. In this way we call attention to points of interaction among the themes of each domain by reference to the categories they share.

As we shall now see, FA captures the structure of PFNET to a very satisfactory degree, and vice versa. This comes as no surprise, since the FA works with the categories having a higher factor loading, and PFNET with those of higher levels of cocitation. FA is good at identifying the thematic areas we would call "consolidated", which are highly interconnected or prominent. Contrariwise, it runs into problems in detecting the least prominent categories or the less consolidated thematic areas; it even identifies some of these as special fields within the same thematic area. Meanwhile, PFNET, with its clusters, graphically displays the possible thematic areas of a scientogram, identifying the most salient category in each, and indicating the route of connection of some thematic areas with others through the sequence of categories that connect them. Yet it does not let us know if a category is prominent or not, or help define the boundaries of the thematic area it comes to create, or its possible denomination. For these reasons, we consider PFNET and FA to be complementary components for the detection of the structure of a scientific domain, as the advantages of one make up for the weaknesses of the other. Thus, FA is in charge of identifying, delimiting, and denominating the thematic areas that make up a scientogram, though it sometimes mistakenly identifies special fields. On the other hand, PFNET will have to make the thematic areas more visible, grouping their categories into bunches, and showing the paths that connect the different prominent categories, and, finally, the different thematic areas of the scientogram of a domain. The categories undetected by any factor, even if they form part of a bunch, will be marked in a color different than the rest of the factors and will not be taken into account in the analysis of the domain.

To evidence the degree of structural fit between the scientogram and the FA, we confirm the degree of fit between factors and category bunches. Likewise, we detect the possible dispersion of categories of one single thematic area throughout the scientogram. Finally, we distinguish the multithematic categories – the red ones – within each thematic area.

Biomedicine

In purple, this thematic area takes up all the central zone of the scientogram, and even much of the upper central section. The categories identified by this factor practically coincide with the bunch detected in the center of the scientogram, and so we could say that in this thematic area we see a nearly complete coincidence between FA and the tandem of raw category cocitation plus PFNET. We use the expression "nearly complete" with caution, because this central grouping includes two categories that could not be identified by the factor – dark gray in color – and are directly connected to others that indeed were factorized. This is the case of *Rehabilitation*, connected to *Sport Sciences*, and situated in the upper left part of the factor; and of *Material Sciences, Biomaterials*, linked to *Engineering Biomedical* and situated in the lower right margin of that space.

In the peripheral zones of this large area we can find borders with other thematic areas. In fact, small thematic areas are incrusted within the upper and upper right edges, while disperse categories join other areas:

- The area of *Orthopedics* touches on the disciplines *Emergency Medicine & Critical Care* and *Otorhinolaringology*.
- The thematic area identified as *Animal Biology and Ecology* is interconnected with *Zoology* and *Entomology*.
- *Medicinal Chemistry* is also identified as a category in the area *Chemistry*.

Psychology

Emerald green in color, it can be found in one of the bunches in the lower part of the scientogram. At first there appears to be no real coincidence between the categories identified by the factor and the concentration of most of these in the scientogram. This might be interpreted as a slight discrepancy as far as the classification goes, but further consideration shows this not to be the case. The reason they do not all appear in the same location is that some belong to other thematic areas. As we shall see below, and in view of the impossibility to be two places at the same time, these end up situated in areas where their pertinence to a thematic area is stronger. Nonetheless, there are two categories that appear dislocated with respect to the main group, specifically in the upper left part of *Biomedicine* area, and they are *Special Education* and *Women´s Studies*. As far as the points of interaction are concerned,

- *Social Sciences, Biomedical* belong as well to the area *Health Policy and Medical Services*.
- *Mathematical Psychology* is identified obviously in the area of *Applied Mathematics* as well.
- *Social Issues, Communication* and *Social Sciences Interdisciplinary* are also included as categories in the area of *Management, Law and Economics*.
- *Experimental Psychology* shares space in the area of *Etiology*.

Materials Sciences and Applied Physics

These, shown in peach color, extend over the right part of the scientogram, along its main branch. The coincidence between the categories identified by FA and their situation on the scientogram is total. There are no categories of this thematic area embedded in others. Yet on the branch where they are distributed, we can see in the scientogram the forking off of other, much smaller branches. In many cases, both the FA and the Cocitation/PFNET tandem identify these as new thematic areas. There are again some categories with double thematic ascription, shown in red:

- *Chemical Engineering, Crystallography,* and *Polymer Science,* identified as well in the area of *Chemistry.*
- *Mechanics* and *Material Sciences Characterization & Testing,* deemed also fit for the area of *Mechanical Engineering.*
- *Optics,* selected as belonging likewise to the area of *Nuclear and Particle Physics.*

Earth and Space Sciences

In light green, they occupy the lower central sector of the scientogram, taking up most of the cluster in the right part of this zone. There is fairly complete coincidence between the number of categories identified by FA and their situation in the scientogram. But one category in the upper mid-right section, *Engineering Petroleum,* appears on its own, connected with a gray mate that has not been identified by any factor. This is a logical connection, on the other hand: around the categories identified by FA and placed in the same bunch, we see other dark gray categories such as *Environmental Sciences, Engineering Environmental, Water Resources, Engineering Civil, Engineering Marine,* and *Imaging Science & Photographic Technology,* which might be included as full members, as they were not taken into account for any other thematic area, but left near this bunch. In light of the methodology we recommend, and the fact that FA did not identify them, we will not consider them parts of this thematic area.

Management, Law and Economics.

Set in the upper left part of the scientogram, in the second highest bunch, and colored in light purple, this thematic area is distributed practically all over this grouping, with the exception of *Urban Studies and Environmental Studies,* which are left out in the colder atmosphere of *Earth and Space Sciences* – lower right section.

From the standpoint of the multiple membership of categories to distinct thematic areas, some are no more than the correspondence with other areas already studied, whereas others may have a very important role in interaction. Such is the case of:

- *Social Issues, Communications and Social Sciences Interdisciplinary.* These are also categories in the area of *Psychology*, as we have seen.
- *Business Finance.* Identified likewise in the area of *Applied Mathematics.*
- *History*, assigned as well to the area of *Humanities,* has a key role in maintaining the integrity of the bunch that constitutes this area. It is furthermore the point of connection with the area of *Humanities,* making it the true point of information interchange between these two thematic areas.
- *Economics.* It is shared with the area of *Applied Mathematics*, and by virtue of its situation and the number of links it has, can be held responsible for the integrity of the grouping in the area of *Management, Law and Economics.* Bearing in mind that it unites these two areas, its importance is evident.

Computer Sciences and Telecommunications

This realm, presented in fuscia, stretches out over a branching in the upper right zone of the scientogram. There is no real discrepancy between the classification proposed by the FA and that obtained by the raw-cocitation/PFNET tandem. Agreement exists to such an extent that even the category *Computer Science Interdisciplinary Applications* (in dark gray), ignored by FA, takes a position in a different bunch than the one occupied by the rest – it can be seen in the right central section of the scientogram.

Agricultural and Soil Sciences

Situated practically altogether in the middle section of the scientogram, this area depicted in the color of green grass shows almost total coincidence with regard to the categories identified by FA and their grouping at some place in the scientogram. Yet there is one category, *Forestry*, in the lower left-center of the image, that usurps the place of *Animal Biology and Ecology*. We underline the fact that this category is connected, in turn, with *Materials Science, Paper and Wood,* and so forestry's incursion in this second thematic terrain is not absurd. Generally speaking, whether the position of a given category seems adequate or not, we should simply remember that the positioning is determined by the dual labor of the raw category cocitation/PFNET tandem.

Humanities

Blue in the face, so to speak, these are situated in the upper left section of the scientogram, within the first bunch of this zone. The coincidence between the FA categories and the RCC/PFNET tandem is quite precise. Only one category – *Philosophy* – can be seen on its own, toward the upper center of the gram, just above *Biomedicine*.

At the same time, just one category has multithematic status: *History*, which as we saw previously, is shared with the thematic area of *Management, Law and Economics*.

Chemistry

Dark brown in color, this is situated essentially toward the right-center, all around the category *Multidisciplinary Chemistry*, which is shown in peach. Of the eight categories that comprise it, four stray from the central bunch: *Material Sciences, Textiles* and *Chemistry, Applied* are seen in the lower left center of the scientogram, connected specifically to *Food Sciences & Technology*, from the area of *Biomedicine*. The other two are multithematic categories: *Medicinal Chemistry* is connected to *Biomedicine*; whereas *Chemical Engineering* shares knowledge with the area of *Materials Science, Applied Physics*.

Etiology

This holds three categories within: two dark green in color, and one red. The first two roam in the central area of the scientogram, as if they were an appendix to the category *Neurosciences*, as part of *Biomedicine*. The red, multithematic category *Experimental Psychology* takes position toward the lower mid-section of the scientogram, in the company of *Psychology*.

We might suspect that the limited number of categories and low factor loading of *Etiology* would make it virtually disappear or disintegrate into another subject area. Verification of such a hypothesis is impossible from viewing the scientogram of just 1 year, which brings us to the usefulness of evolutive scientograms, to confirm the temporal sequence of categories of interest.

Mechanical Engineering

Around the category *Material Sciences, Multidisciplinary*, in the right half of the scientogram, we can spot this field, depicted in yellow. Two of its eight constituent categories appear in red. *Mechanics and Material Sciences Characterization & Testing* are shared by this thematic area and by that of *Material Sciences and Applied Physics*. And two others are dispersed over the left central area of the scientogram: *Engineering* is linked to *Applied Mathematics*, and *Acoustics* is connected to *Radiology, Nuclear Medicine and Medical Imaging*, of the area *Biomedicine*.

Health Policy, Medical Services.

Deep-sea green in shading, these six categories float above *Biomedicine*, toward the top of the display. This thematic area has two branches that derive from *General and Internal Medicine*. One is connected directly, the other indirectly, through the category *Public, Environmental, and Occupational Health*. Therefore, this category seems a bit isolated, above all if we bear in mind that one of its component categories – *Medical Informatics* – is situated at the left center of the scientogram in the space mostly occupied by *Applied Mathematics*, and connected to the category *Statistics and Probability*.

As we saw earlier, this area, along with *Psychology*, contributes to the category *Social Sciences, Biomedical*, and therefore shares its sources as well.

Applied Mathematics

This gray matter can be seen in the left-center of the scientogram, where its main branch forks gently into two directions. Interestingly enough, one of the categories of this area – *Mathematics, Miscellaneous* – is the one that appears to give rise to the two resulting branches. This means it is an important element for the analysis and interpretation of the scientogram more on that is discussed later. The coincidence between the categories identified by FA and by the working tandem is complete, and deserves our attention.

Of the two branches that come out of this category, the one toward the lower left shows us how a series of unfactorized categories (in gray) dangles from the branch. In our opinion, *Mathematics* should form part of this thematic area, although it was not identified by FA. The same could be said of the rest: *Business Management, Engineering Manufacturing, Engineering Industrial*, and *Operation Research & Management Science*, as they themselves constitute a branch from the latter, which in turn is connected to *Applied Mathematics*. Although the contents of the first four is not strictly mathematical, they all converge in *Operation Research & Management Science*, whose objective is the scientific or mathematical analysis of problems related with complex systems. We will uphold the proposed methodology, even though it does not support our opinion in this case.

The seven categories identified by FA include three that are multithematic:

1. *Psychology, Mathematical* which, as we saw earlier, shares with the area of *Psychology*, but is situated here because its relationship is much stronger with the category *Mathematics, Applied* than with *Psychology*.
2. Business Finance, identified also in the area of Management, Law and Economics.
3. *Economics*. Shared by the areas of *Applied Mathematics* and *Management, Law and Economics*, it evidences a genuine point of friction, and of informational exchange, between the two areas mentioned.

Nuclear and Particle Physics

Mauve in color, it occupies the right central area of the scientogram, forming a small bunch around the category *Physics Multidisciplinary*, of the area *Material Science and Applied Physics*. Only one of the categories identified by the FA appears separate from the bunch formed by the rest: Education Scientific Disciplines, which is at the upper midsection of the display, just above the area of *Biomedicine*, where some of the disjointed categories of its thematic groupings linger.

As for the multithematic categories, we have *Optics*, not surprising in view of its shared terrain with the area *Materials Sciences and Applied Physics*, whence it comes.

Animal Biology, Ecology

It is situated in the left center of the scientogram and appears in yellow. This area has as its point of departure the category *Biology, Miscellaneous*, in red, as it belongs to *Biomedicine* as well. There are also multithematic categories that likewise pertain to *Biomedicine*: for instance, *Entomology and Zoology*, placed in the center of the scientogram because of their greater affinity with this thematic area. Of the eight categories identified by FA, the geographical coincidence with the PFNET bunches is complete, except for the two categories already mentioned.

Orthopedics

Consisting of three categories and situated, in salmon, at the center of the visualization, it has only one visible category in this color, *Orthopedics*, since *Otorhinolaryngology* and *Emergency Medicine & Critical Care* are multithematic areas shared with *Biomedicine* and therefore depicted in red. There is scarcely any coincidence between FA and the tandem: *Orthopedics and Otorhinolaryngology* link with the category *Surgery*, of the area of *Biomedicine*; and *Emergency Medicine & Critical Care* connect with the category *Medicine General & Internal*, also of the area of *Biomedicine*. This lack of coincidence, together with the low number of categories that integrate it and the fact that two of them are multithematic, makes us doubt whether this factor in itself is a special field of *Biomedicine* or rather a new thematic area in the process of development or disappearance. Faced with the impossibility of confirming either (with no scientograms from years before or afterward), and ever-faithful to our methodology, we recognize it as an area, and not as a specialized field.

7.1.3 Analysis of a Domain

Bearing in mind that our scientograms are extreme schematizations of the scientific output of a domain, and that their structure is determined by the tandem raw category cocitation plus PFNET, their analysis and interpretation will be based on the inferences that can be derived from the resulting PFNET structure. That is, of the categories with a high degree of interconnection, of the principle of triangular inequality, and of the paths with a greater weight or importance.

For instance, if a category or thematic area occupies a central position in the scientogram, it is very likely that one or the other have a more general or universal nature in the domain as the consequence of the number of sources they share with the rest and that, therefore, contribute most to the development of that area. Accordingly, in general, the more peripheral the situation of a category, the more exclusive its nature, the fewer sources it will share with other categories, and the lesser its contribution to advancements in that area. Occupying a central position is also important from the point of view of communication, as thanks to this situation, it is possible for other categories or thematic areas to be interconnected through it. This phenomenon would not occur if the given area or central category did not exist. If, for example, the thematic area of *Biomedicine* disappeared altogether from the scientogram of Fig. 7.2, the rest of the areas would be totally disconnected from each other. Something very similar would happen with the other categories if *Biochemistry & Molecular Biology* were eliminated. We can say the same of a number of other areas or categories, although the loss of interconnection will be less severe if the positions involved are not as central. This manner of analyzing and interpreting the scientograms does not only explain the patterns of cocitation implicit in a domain, but also provides an intuitive explanation, practical for experts and non-experts, of the inner workings of PFNET.

Macrostructure

The macrostructural analysis of a domain will be carried out with the corresponding factor scientogram. To this aim, we shall resort to each one of the thematic areas identified in it and proceed to their interpretation in agreement with the criterion mentioned earlier. Yet in this case, in order to secure a more compact vision of the intellectual structure of the domain and facilitate its analysis, we shall build a new scientogram based on the graphic representation of the centroid of the categories, with a value equal to or greater than 0.5 (Salton and McGill, 1983), of each one of the 16 factors or thematic areas that the FA identifies in the original cocitation matrix of 218 x 218 categories. The result is given below.

The scientogram of world output 2002 in Fig. 7.3 consists of 16 thematic areas, as did the factor display of Fig. 7.2. And as in the scientograms shown thus far, the volume is shown proportional to the volume of documents represented.

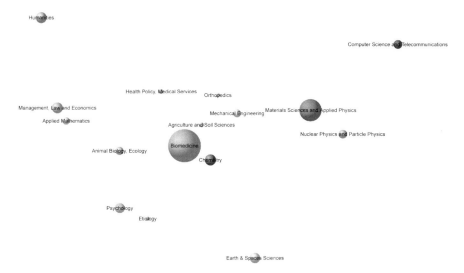

Fig. 7.3. Scientogram of centroids of the categories that make up the thematic area of world output 2002

The first thing to grab our attention with this new scientogram is the combination of just a few thematic areas of considerable size, along with other very small ones. They thereby reflect the hyperbolic nature of all bibliometric distributions (Small and Garfield, 1985). Another noteworthy aspect is the central–peripheral pattern, where a major central subject area serves as the node of connection to other smaller ones around it.

In a broad sense, we can say that in the central zone, we see what could be called *Biomedical and Earth Sciences: Biomedicine, Psychology, Etiology, Animal Biology and Ecology, Health Policy, Orthopedics, Earth and Space Science, and Agriculture and Earth Sciences*. To the right appear the "hard" Sciences: *Material Sciences and Applied Physics, Engineering, Computers and Telecommunications, Nuclear and Particle Physics, and Chemistry,* and to the left, the softer sciences, such as *Applied Mathematics, Management, Law and Economics, and Humanities*. This scheme of macrostructural vertebration of the sciences is a typical and persistent device in the scientograms of developed countries, but not in still-developing countries, as we have seen in other scientographic work.

The FA is decidedly adept at capturing the structure of PFNET, and vice versa, as we mentioned in an earlier section. If we compare the positions of the thematic areas of the scientogram of Fig. 7.3 with the respective positions in the scientogram of Fig. 7.2, we cannot ignore the astounding similarity. Moreover, if in the scientogram of centroids we reproduce the paths or links connecting the different subject areas of the factor scientogram, we obtain a reduced, compact graphic representation of the latter. This fact will facilitate interpretation.

For analysis of the scientograms (Fig. 7.4), and from the PFNET standpoint, we will apply two measures: centrality and prominence. A thematic area is said to be central if its nodal grade is high with respect to the rest. On the other hand, it is prominent when its links make it particularly visible as opposed to the rest. By generating a network with as many nodes as there are thematic areas, and manually reproducing the connection between these areas with links, using *Pajek* (Batagelj and Mrvar, 1998), we can easily calculate these measures.

Centrality

The PFNET perspective leads us to state that on a world level in the year 2002, the most central thematic area is *Biomedicine*, by virtue of its higher nodal grade (Table 7.10).

This means that *Biomedicine* is the most universal thematic area, with more shared sources, and the one most involved in the domain's development. Wielding a much lesser production, but also quite central (Fig. 7.3) are

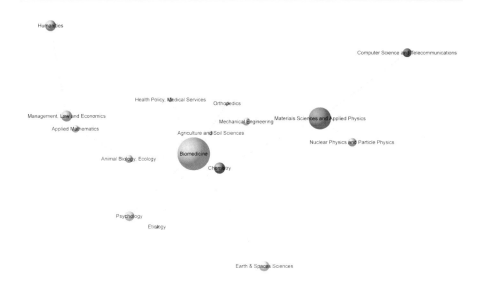

Fig. 7.4. Scientogram of the thematic areas of world science, 2002

Table 7.10. Centrality of grade of the thematic areas of world scientific output.

Thematic area	Grade
Biomedicine	8
Materials Sciences and Applied Physics	4
Animal Biology, Ecology	2
Management, Law and Economics	2
Applied Mathematics	2
Chemistry	2
Agriculture and Soil Sciences	1
Earth & Space Sciences	1
Etiology	1
Nuclear Physics and Particle Physics	1
Humanities	1
Computer Science and Telecommunications	1
Mechanical Engineering	1
Orthopedics	1
Health Policy, Medical Services	1
Psychology	1

Agriculture and Earth Science, and *Chemistry*. This is almost to be expected, as in the real world, agriculture is the economic foundation of many countries, while chemistry is an essential element in medicine, the food sector, industry, and more. The most peripheral areas, which share fewer sources for domain developments, are the *Humanities, Computer Science and Telecommunications,* and *Earth and Space Sciences.* The other subject areas, such as *Psychology, Animal Biology* and *Ecology, Materials Sciences and Applied Physics*, and so on, occupy more or less intermediate positions.

Prominence

To detect the outstanding or most prominent theme areas of a domain, we propose the use of a "robbery" algorithm, included in the Pajek computer program, through which the strongest nodes, or those with a higher grade, steal links or degree from their weakest neighbors. The result is a vector that indicates the level of prominence of the different thematic areas at the same time that it allows us to modify their size, in order to make them more visible in the scientogram, as shown by Table 7.11 and Fig. 7.5.

If we compare the results obtained with centrality with those arrived at through prominence, we see there is little overlap. This is how things should be: the first establishes criteria to determine the level of universality of shared sources; whereas the second constitutes the order of prominence

Table 7.11. Prominence of the thematic areas of the world

Thematic area	Grade
Biomedicine	8.5
Materials Sciences and Applied Physics	4.5
Management, Law and Economics	2
Applied Mathematics	1
Agriculture and Soil Sciences	0
Animal Biology, Ecology	0
Earth & Space Sciences	0
Etiology	0
Nuclear Physics and Particle Physics	0
Humanities	0
Computer Science and Telecommunications	0
Mechanical Engineering	0
Orthopedics	0
Health Policy, Medical Services	0
Psychology	0
Chemistry	0

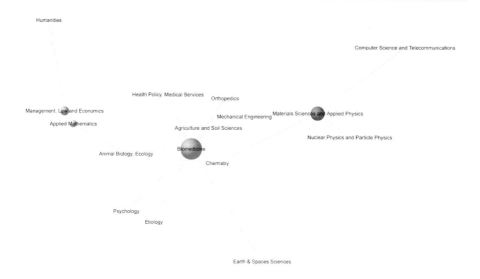

Fig. 7.5. Most prominent thematic areas of world science, 2002

in terms of the relations that each subject area maintains with the others. Consequently, where they do indeed coincide is in pointing to the area of *Biomedicine* as the most central and most prominent of the domain.

Microstructure

In order to carry out a microstructural analysis of a domain, we may resort either to its basic scientogram, which shows the semantic structure of the domain, or else to its factor scientogram. The latter reveals the intellectual structure; yet as a rule of thumb, it will always be preferable to use the factor scientogram, as its thematic groupings let us identify, at a glance, the most prominent categories or their origins. These are characterized by their high degree of interconnectivity, or by serving as structural hubs. To go on with this analysis, we refer to the scientogram of Fig. 7.2.

The scientogram of world scientific output for 2002 consists of 218 categories and 217 links that bind them together. None of the links is isolated. As we saw earlier, in the case of subject area, what most attracts our attention in this scientogram is the existence of a handful of large categories – with considerable output – and many of a smaller size, showing again the hyperbolic nature of all bibliometric distributions (Small and Garfield, 1985). The small size can be seen above all in the center and right center of the scientogram, and to a lesser extent in the left half. That is, there is a greater production on the part of the categories pertaining to the hard sciences and medical sciences than to those linked to the softer sciences. The second most interesting aspect is that the pattern of connections that the categories adopt is of the central–peripheral type: we find a large central category that serves as the hub of interconnection with other categories around it, while responsible for maintaining the structural cohesion of the scientogram.

Centrality

The naked eye has no doubt in this case: the most central category of the scientogram of world scientific output for 2002 is *Biochemistry & Molecular Biology*. The centrality of degree of the depiction comes to confirm this fact (Table 7.12).

The maximal centrality of *Biochemistry & Molecular Biology* crowns it as the most general and universal of the 218 categories that contributed to world scientific publication in the year 2002. It is the category with the most shared sources, the top contributor to world scientific advancement, and so it merits its place at the core of the visual representation. This position enhances its connectivity and the interchange of knowledge with or among the greatest number of categories, making it the central axis of the vertebration of science. If we were to eliminate *Biochemistry and Molecular Biology*, the surrounding categories would be left totally disconnected, as would the semantic structure of the scientogram.

Table 7.12. Top 16 categories of the highest world grade

Category	Grade
Biochemistry & Molecular Biology	31
Psychology	10
Medicine, General & Internal	9
Materials Science, Multidisciplinary	9
Economics	9
Geosciences, Interdisciplinary	8
Chemistry, Multidisciplinary	6
Environmental Sciences	6
Public, Environmental & Occupational Health	6
Physics, Multidisciplinary	5
Psychiatry	5
Immunology	5
Engineering, Electrical & Electronic	5
Political Science	5
History	5
Literature	5

Having identified and crowned the most central category of our domain, the order of the rest will be directly related with the distance of each from the core. To establish this ranking of category centrality, we refer to the geodesic distance – the shortest path between two nodes of a network – existing between any given category and *Biochemistry & Molecular Biology* (Table 7.13).

Table 7.13. Distances of the categories of the world scientogram with respect to the central category

Category	Dist.	Category	Dist.
Biochemistry & Molecular Biology	0	Psychology	4
Medicine, General & Internal	1	Limnology	4
Entomology	1	Ornithology	4
Chemistry, Multidisciplinary	1	Psychology, Mathematical	4
Nutrition & Dietetics	1	Materials Science, Characterization & Testing	4
Plant Sciences	1	Mining & Mineral Processing	4
Zoology	1	Environmental Studies	4
Biology	1	Psychology, Clinical	4
Biotechnology & Applied Microbiology	1	Materials Science, Textiles	4
Biochemical Research Methods	1	Social Sciences, Mathematical Methods	4
Cell Biology	1	Education, Special	4
Dermatology & Venereal Diseases	1	Archeology	4
Endocrinology & Metabolism	1	Psychology, Psicoanálisis	4
Hematology	1	Philosophy	4
Medicine, Research & Experimental	1	Geosciences, Interdisciplinary	5
Neurosciences	1	Engineering, Petroleum	5
Pathology	1	Medical Informatics	5
Oncology	1	Mathematics, Applied	5
Pharmacology & Pharmacy	1	Engineering, Aerospace	5
Physiology	1	Physics, Multidisciplinary	5
Parasitology	1	Physics, Applied	5
Virology	1	Engineering, Mechanical	5
Microbiology	1	Materials Science, Paper & Wood	5
Biophysics	1	Ergonomics	5
Medical Laboratory Technology	1	Fisheries	5
Genetics & Heredity	1	Psychology, Experimental	5
Immunology	1	Engineering, Civil	5

Table 7.13. (Cont.)

Category	Dist.	Category	Dist.
Geriatrics & Gerontology	1	Psychology, Developmental	5
Urology & Nephrology	1	Education & Educational Research	5
Developmental Biology	1	Psychology, Educational	5
Mycology	1	Economics	5
Microscopy	1	Urban Studies	5
Gastroenterology & Hepatology	2	Social Sciences, Interdisciplinary	5
Radiology, Nuclear Medicine & Medical Imaging	2	Psychology, Social	5
Agriculture	2	Criminology & Penology	5
Food Science & Technology	2	Family Studies	5
Anatomy & Morphology	2	Psychology, Applied	5
Surgery	2	Architecture	5
Crystallography	2	Mathematics	6
Clinical Neurology	2	Astronomy & Astrophysics	6
Obstetrics & Gynecology	2	Physics, Nuclear	6
Ophthalmology	2	Physics, Atomic, Molecular & Chemical	6
Pediatrics	2	Engineering, Electrical & Electronic	6
Polymer Science	2	Engineering	6
Veterinary Sciences	2	Agricultural Economics & Policy	6
Chemistry, Organic	2	Paleontology	6
Chemistry, Analytical	2	Instruments & Instrumentation	6
Chemistry, Physical	2	Geochemistry & Geophysics	6
Chemistry, Inorganic & Nuclear	2	Oceanography	6
Behavioral Sciences	2	Physics, Particles & Fields	6
Toxicology	2	Operations Research & Management Science	6
Infectious Diseases	2	Engineering, Marine	6
Rheumatology	2	Thermodynamics	6
Allergy	2	Materials Science, Coatings & Films	6
Biology, Miscellaneous	2	Geography	6
Public, Environmental & Occupational Health	2	Geology	6
Emergency Medicine & Critical Care	2	Physics, Mathematical	6
Medicine, Legal	2	Sociology	6
Peripheral Vascular Disease	2	Law	6
Sport Sciences	2	Planning & Development	6
Engineering, Biomedical	2	Business, Finance	6

Table 7.13. (Cont.)

Category	Dist.	Category	Dist.
Chemistry, Medicinal	2	Social Work	6
Horticulture	2	Political Science	6
Transplantation	2	History Of Social Sciences	6
Health Care Sciences & Services	2	Industrial Relations & Labor	6
Education, Scientific Disciplines	3	Engineering, Geological	6
Materials Science, Multidisciplinary	3	Optics	7
Acoustics	3	Automation & Control Systems	7
Cardiac & Cardiovascular Systems	3	Computer Science, Artificial Intelligence	7
Agriculture, Dairy & Animal Science	3	Telecommunications	7
Agriculture, Soil Science	3	Mineralogy	7
Anesthesiology	3	Computer Science, Hardware & Architecture	7
Environmental Sciences	3	Nuclear Science & Technology	7
Dentistry, Oral Surgery & Medicine	3	Physics, Fluids & Plasmas	7
Ecology	3	Spectroscopy	7
Orthopedics	3	Computer Science, Interdisciplinary Applications	7
Otorhinolaryngology	3	Electrochemistry	7
Engineering, Chemical	3	Engineering, Industrial	7
Psychiatry	3	Remote Sensing	7
Tropical Medicine	3	Management	7
Reproductive Biology	3	Social Issues	7
History & Philosophy Of Science	3	Public Administration	7
Anthropology	3	Area Studies	7
Rehabilitation	3	International Relations	7
Chemistry, Applied	3	History	7
Mathematics, Miscellaneous	3	Ethnic Studies	7
Materials Science, Biomaterials	3	Communication	7
Social Sciences, Biomedical	3	Demography	7
Psychology, Biological	3	Engineering, Ocean	7
Transportation	3	Computer Science, Theory & Methods	8
Nursing	3	Engineering, Manufacturing	8
Women's Studies	3	Computer Science, Cybernetics	8
Health Policy & Services	3	Imaging Science & Photographic Technology	8
Energy & Fuels	4	Business	8

Table 7.13. (Cont.)

Category	Dist.	Category	Dist.
Construction & Building Technology	4	Literary Reviews	8
Respiratory System	4	Music	8
Water Resources	4	Literature	8
Marine & Freshwater Biology	4	Arts & Humanities, General	8
Metallurgy & Metallurgical Engineering	4	Religion	8
Mechanics	4	Computer Science, Software, Graphics, Programming	9
Substance Abuse	4	Language & Linguistics	9
Statistics & Probability	4	Theater	9
Materials Science, Composites	4	Poetry	9
Physics, Condensed Matter	4	Literature, American	9
Meteorology & Atmospheric Sciences	4	Art	9
Forestry	4	Computer Science, Information Systems	10
Materials Science, Ceramics	4	Literature, Romance	10
Engineering, Environmental	4	Literature, Slavic	10
Andrology	4	Information Science & Library Science	11

Translating these distances to a vector, and giving a distinctive color to each, leads us to a new scientogram that quickly and easily informs the eye of the proximity or distance of each category with respect to the central one, as seen in Fig. 7.6. To facilitate the calculation of distances, the lower left corner of the scientogram contains a table of equivalences between color and distances.

Prominence

Another of the characteristics of the raw category cocitation plus PFNET tandem is the joint potential in identifying the most salient categories of a domain. The detection of these categories is important from an analytical point of view, as they are usually the nuclei of the thematic areas.

Again basing our comments on the scientogram of Fig. 3.2, and at plain sight, we quickly spot the most prominent categories. For instance, it is obvious that the most prominent of all is the central category – *Biochemistry and Molecular Biology*. From that position, going clockwise from twelve o'clock, we see just above this category that of *Medicine General and Internal Medicine*, and dependent on it, the category *Public Environmental & Occupational Health*. Toward a quarter past three we see *Chemistry*

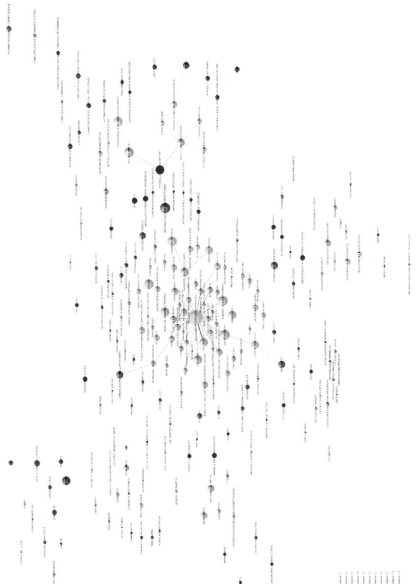

Fig. 7.6. Scientogram of world differences (in scientific output), with respect to the central category

Multidisciplinary, Material Science Multidisciplinary, Electrical Engineering & Electronics, and *Physics Multidisciplinary*. Around half past six on our hypothetical clock we encounter *Environmental Sciences* and, hanging from it, *Geosciences Interdisciplinary*. A bit to the left is *Immunology*, as well as *Psychiatry*, and from the latter dangles *Psychology*. Near a quarter to ten we have *Economics* and *History* as the prominent categories.

Out of the 14 categories detected at a glance, 10 form the source or nucleus of a thematic area: *Biochemistry & Molecular Biology, Public Environmental and Occupational Health, Chemistry Multidisciplinary, Material Science Multidisciplinary, Engineering Electrical and Electronic, Physics Multidisciplinary, Geosciences Interdisciplinary, Psychiatry, Economics*, and *History*.

This visual method used by *Chen* and *Carr* (1999b) for the localization of specialized fields in author cocitation PFNET networks also gives good results in the detection of thematic areas in PFNET networks of category cocitation, as we have just demonstrated ... above all, when one knows what he or she is looking for. In our case, for instance, we decided that a category was prominent if, in it, four or more links converged. This clue led us to the source of ten thematic areas.

Other authors such as *White* (2003) use the measures of grade centrality to create a vector, which is then used to modify the size of the nodes of a PFNET, and so to rapidly detect the most prominent ones. Although this alternative method can prove very useful in author cocitation PFNET to discover specialized fields, it is not very appropriate for category cocitation, because more of agglomerations take place (though to a lesser degree), than with the author circumstances. Hence, and to ensure that the number of categories detected visually is not too high, the nodes must be made very small in size – yet always proportional with measures of grade centrality – which interferes with visual detection. At any rate, all the categories that are the beginning point of a thematic area are detected, even when the number of prominent elements is excessive. Fig. 7.7 offers proof of this.

Our research team, as in the previous thematic areas, resorted to the robbery algorithm to detect and make easily visible the most prominent categories. The result can be observed in the following vector (Table 7.14) and in the scientogram of Fig. 7.8.

Of the 29 categories detected by the algorithm as prominent ones, we identify the first 15 as the most representative using the scree test. Of these 15, 12 coincide with the origin of a thematic area: *Biochemistry & Molecular Biology, Material Science Multidisciplinary, Engineering Mechanical, Engineering Electrical and Electronics, Physics Multidisciplinary*,

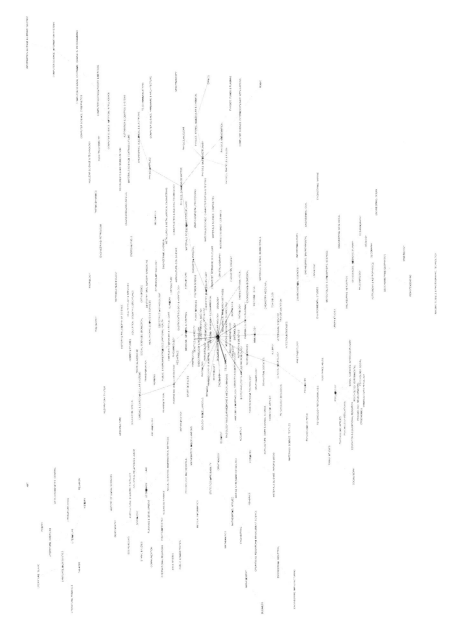

Fig. 7.7. Scientogram of grade centrality for world scientific output 2002

Table 7.14. Prominence of the top 16 categories of the world

Category	Prominence
Biochemistry & Molecular Biology	152.04
Economics	32.67
Psychology	30.22
Geosciences, Interdisciplinary	25.29
Engineering, Electrical & Electronic	24.67
Materials Science, Multidisciplinary	21.50
Physics, Multidisciplinary	19.83
Mathematics, Applied	17.00
Environmental Sciences	16.05
Literature	13.00
Ecology	11.00
History	10.71
Mathematics, Miscellaneous	7.62
Food Science & Technology	7.18
Biology, Miscellaneous	6.23
Energy & Fuels	3
Radiology, Nuclear Medicine & Medical Imaging	3
Computer Science, Information Systems	3
Cardiac & Cardiovascular Systems	3
Engineering, Mechanical	3
Reproductive Biology	3
Rehabilitation	3
Engineering, Biomedical	3
Engineering, Civil	3
Archeology	3
Family Studies	3
Computer Science, Software, Graphics, Programming	2
Peripheral Vascular Disease	2
Sport Sciences	2

Geosciences Interdisciplinary, Psychiatry, Food Sciences & Technology, Biology Miscellaneous, Mathematics Miscellaneous, Economics, and *History*, as shown in Fig. 7.8.

This method combines the strong suits of the two previous ones: it affords the reliability and precision of the first, while also offering all the advantages of a visual detection system. In this way, and for the case at hand, 12 of the 16 thematic areas detected by FA are rapidly identified, whose origin or nucleus is a prominent category. True, another three are mistakenly identified – they are not really the source of any subject area, in agreement with this new method. In terms of the thematic nature of the categories detected, the method allows us to distinguish three types of prominent categories:

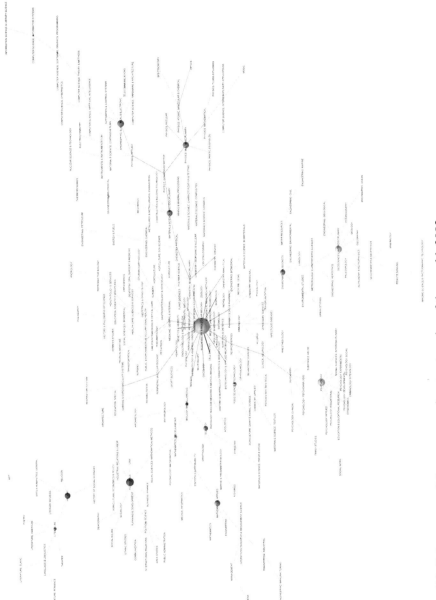

Fig. 7.8. Scientogram of the most prominent categories of the world, 2002

1. Those that make manifest the imprecision of the borders between the different thematic areas that constitute a single domain. Making manifest the multidisciplinary character of the sources that constitute these categories. Such is the case of *Economics, History, and Biology Miscellaneous*. For example, three major branches come out of *Economics*, indicating its degree of multidisciplinarity; they are *Political Sciences, Sociology*, and *History of Social Sciences*. From *History*, in turn, sprout two branches with a weight of their own: *Literature and Linguistics*, on the one hand, and *Art and Humanities*, on the other. And from *Biology, Miscellaneous*, two branches appear, one going downward that we could identify as *Animal Biology and Ecology*, and the upward-branching beginnings of *Human Biology*. On some occasions, these categories mark the origin or the appearance of special fields, as does *Mechanics*. This category indicates the point of convergence of what FA identifies as two thematic areas: *Materials Science and Applied Physics*, and *Mechanical Engineering*. Such salient categories mark the origin or initial phase in the evolution of a thematic area.

2. Those that belong to an area other than the one they give rise to. They are the clear borderline defined between two thematic areas. While integrated and consolidated in one theme area, they are the main point of interchange of knowledge among the different areas that they put into contact. In our case, situated in the upper-middle and upper-right part of the scientogram are *Public Environmental & Occupational Health, Chemistry Multidisciplinary, Material Sciences Multidisciplinary, Engineering Mechanical, Engineering Electrical & Electronic*, and *Physics Multidisciplinary*. This type of prominent category signals well-established special fields as well as thematic areas in developing stages.

3. Those that belong to the same thematic area that they identify. They denote totally integrated and consolidated categories in that area. They are the source of common knowledge for the rest of the categories making up the area. Located mainly in the lower midsection of the scientogram, they are *Biology & Molecular Biology, Geosciences Interdisciplinary, Psychology*, and *Mathematics Miscellaneous*, each identifying a perfectly consolidated thematic area.

Four are the thematic areas that could not be detected in the scientogram we came up with: Etiology, Health Policy and Medical Services, Agriculture and Soil Sciences, and Orthopedics. This fact also kept us from identifying the nuclei of their corresponding thematic areas. The reason is the reduced number of categories that constitute them – a consequence of the fragmentation that the ISI introduces with its categorization – and the dispersion or lack of concentration around a common category that would make them become prominent. Thus, only FA can identify these thematic areas; a reminder of its importance, and that its prominent categories must be identified by the criterion of the researcher. These four categories are *Behavioural Science, Medicine General & Internal, Plants Science,* and *Surgery.* If we look closely, we see that these four categories are among the ones we denominated as type 2, that is, those constituting one subject area while giving rise to another. This means that they may in fact be consolidated specialized areas, or else areas in the process of consolidation or even of gradual disappearance. It is very difficult to determine which, unless we can refer to evolutive scientograms.

To end this section, let us underline the importance of the combination of centers and bunches, as they allow us to identify and confirm the existence of possible thematic areas that go unacknowledged by FA. A case in point is what we might call *Environmental Sciences,* in the lower central section of the scientogram, gray in color, and made up of *Environmental Sciences* as the main category, along with *Environmental Engineering, Water Resources, Engineering Civil* and *Engineering Marine.*

The Backbone of a Domain

One of the benefits of PFNET scientograms based on cocitation is that they provide an x-ray of sorts of the backbone of the research of that domain. They do this through the ability of PFNET to select the most significant links between categories, and because of their graphic capacity to show the intensity of cocitation as the thickness of the links.

To return to the scientogram of world output in 2002, in both its basic and factor versions (Figs. 7.1 and 7.2), we can see how some links uniting sequences of categories are thicker than others. To determine which links and categories are the ones that form the backbone of the domain, we take as reference the highest value of the link that joins two thematic areas and we eliminate the links that remain below that value, along with the disconnected categories. The result is the backbone or the means in which the science of a scientific domain is vertebrated (Fig. 7.9).

The next scientogram we present is stripped down to its backbone in order to study it in clearer detail. Here are the basic thematic areas of the world domain for 2002: *Biomedicine, Materials Sciences and Applied Physics,* and *Chemistry.* Given that *Biochemistry & Molecular Biology* was the category that intervened most in the development of the domain, it is obviously also key here. Curiously, this category shares most resources with its own thematic area, and in second place, with those of other areas.

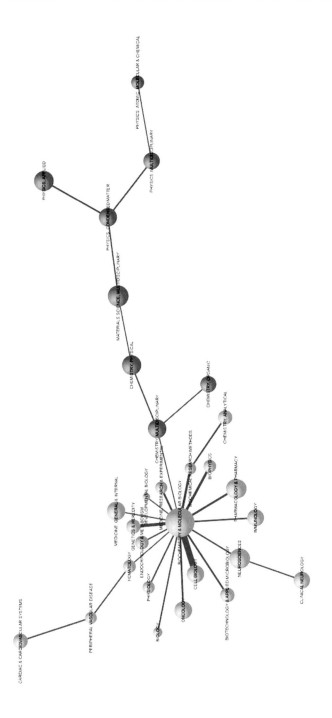

Fig. 7.9. Scientogram of the vertebration of world scientific output 2002

Also remarkable is that despite the extreme simplification offered here, some long distance paths in this area survive, such as *Cardiac & Cardiovascular Systems* ←→ *Peripheral Vascular Disease* ←→ *Hemotology* ←→ *Biochemistry & Molecular Biology*, indicating the extent of multidisciplinarity of these categories.

The area of *Chemistry* appears poorly represented. Yet it is noteworthy that despite its low number of categories in the JCR, it makes its significance known in this bare bones depiction (albeit with a single category – *Chemistry Organic*).

Materials Sciences and Applied Physics appears here as a reduced version of its structural entity in the original scientogram. Notwithstanding, it serves to demonstrate the sequence of its basic structure, which traverses from *Chemistry Multidisciplinary* to *Physics Condensed Matter*, which acts as a bridge over to *Physics Applied*, and to *Physics Atomic Molecular & Chemical*.

7.2 Scientography and Comparison of Domains

In this section, we take on two major geographical domains: the European Union (EU)[5] and the United States of America (US). To this aim, we make a comparative study of their most significant elements under separate headings: the scientogram, FA, the factor scientogram, and so forth, so as to compare these most representative elements and infer similarities and differences.

7.2.1 Scientograms

The following scientogram (Fig. 7.10) is the graphic schematization by PFNET of the 368,120 scientific documents produced by EU member countries in the year 2002[5] and collected in the ISI database, grouped into the 218 categories recognized by the JCR. As in earlier cases, the thickness of the links indicates the intensity of cocitation of the categories they unite, while the size of the spheres reflects the magnitude of production. The point-by-point correspondence between each of the said categories and their output can be derived from the Table 7.15

---

[5] There were 15 EU Member States in 2002: France, Germany, The Netherlands, Belgium, Luxemburg, Italy, Denmark, Ireland, the United Kingdom, Greece, Spain, Portugal, Austria, Finland, and Sweden.

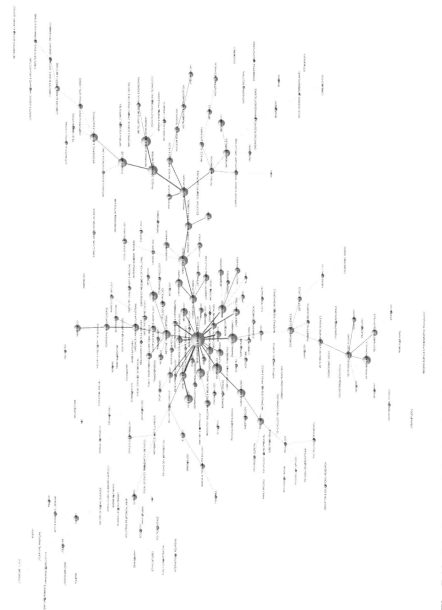

Fig. 7.10. Scientogram of the domain EU scientific output, 2002

Table 7.15. Listing of JCR categories and EU output for 2002

Category	Output	Category	Output
Energy & Fuels	1462	Medicine, Legal	358
Geosciences, Interdisciplinary	5337	Anthropology	630
Engineering, Petroleum	256	Peripheral Vascular Disease	4727
Gastroenterology & Hepatology	4557	Rehabilitation	838
Radiology, Nuclear Medicine & Medical Imaging	4832	Sport Sciences	1677
Mathematics	5356	Instruments & Instrumentation	3367
Medicine, General & Internal	8610	Geochemistry & Geophysics	4636
Education, Scientific Disciplines	365	Mineralogy	810
Medical Informatics	569	Engineering, Environmental	1651
Entomology	1100	Computer Science, Hardware & Architecture	832
Chemistry, Multidisciplinary	6187	Developmental Biology	1566
Construction & Building Technology	608	Andrology	110
Materials Science, Multidisciplinary	10666	Engineering, Biomedical	1984
Computer Science, Theory & Methods	3425	Oceanography	2704
Computer Science, Information Systems	1466	Psychology	1500
Computer Science, Software, Graphics, Programming	1565	Nuclear Science & Technology	3452
Mathematics, Applied	5689	Chemistry, Applied	1863
Acoustics	1209	Physics, Fluids & Plasmas	3033
Cardiac & Cardiovascular Systems	6287	Information Science & Library Science	640
Respiratory System	3110	Physics, Particles & Fields	5024
Agriculture, Dairy & Animal Science	1430	Chemistry, Medicinal	2229
Agriculture	1657	Materials Science, Paper & Wood	421
Agriculture, Soil Science	1084	Ergonomics	273
Nutrition & Dietetics	2445	Spectroscopy	2945
Food Science & Technology	3124	Operations Research & Management Science	1361

Table 7.15. (Cont.)

Category	Output	Category	Output
Anesthesiology	2514	Engineering, Marine	45
Anatomy & Morphology	571	Thermodynamics	1026
Engineering, Aerospace	770	Materials Science, Coatings & Films	1156
Astronomy & Astrophysics	8194	Fisheries	964
Biochemistry & Molecular Biology	23469	Limnology	453
Plant Sciences	5619	Mathematics, Miscellaneous	439
Zoology	2508	Geography	854
Biology	2557	Ornithology	404
Biotechnology & Applied Microbiology	5262	Materials Science, Biomaterials	660
Surgery	8913	Geology	755
Crystallography	2340	Computer Science, Interdisciplinary Applications	2238
Biochemical Research Methods	3728	Horticulture	717
Cell Biology	8671	Computer Science, Cybernetics	294
Dermatology & Venereal Diseases	2898	Transplantation	2083
Endocrinology & Metabolism	6277	Psychology, Experimental	1676
Hematology	5622	Psychology, Mathematical	114
Environmental Sciences	5566	Materials Science, Characterization & Testing	375
Water Resources	2314	Engineering, Civil	1398
Marine & Freshwater Biology	3206	Mining & Mineral Processing	400
Metallurgy & Metallurgical Engineering	2027	Physics, Mathematical	3738
Mechanics	3355	Electrochemistry	1112
Medicine, Research & Experimental	3528	Psychology, Developmental	643
Neurosciences	11944	Education & Educational Research	1072
Clinical Neurology	8467	Engineering, Industrial	701
Pathology	3094	Mycology	540
Obstetrics & Gynecology	3243	Psychology, Educational	310
Dentistry, Oral Surgery & Medicine	1910	Environmental Studies	1173
Ecology	3229	Social Sciences, Biomedical	346

Table 7.15. (Cont.)

Category	Output	Category	Output
Oncology	9253	Psychology, Biological	476
Ophthalmology	2389	Sociology	1285
Orthopedics	1935	Law	397
Otorhinolaryngology	1419	Microscopy	383
Pediatrics	3766	Remote Sensing	463
Pharmacology & Pharmacy	10997	Transportation	128
Physics, Multidisciplinary	8030	Psychology, Clinical	1090
Physiology	4023	Imaging Science & Photographic Technology	372
Polymer Science	3341	Materials Science, Textiles	194
Physics, Applied	8939	Economics	3743
Engineering, Chemical	4179	Social Sciences, Mathematical Methods	554
Psychiatry	5887	Business	779
Tropical Medicine	563	Management	1154
Parasitology	1093	Urban Studies	406
Veterinary Sciences	3778	Social Sciences, Interdisciplinary	796
Virology	2285	Planning & Development	740
Substance Abuse	471	Business, Finance	401
Engineering, Mechanical	2461	Social Issues	308
Microbiology	6609	Public Administration	421
Statistics & Probability	1933	Social Work	365
Physics, Nuclear	3668	Psychology, Social	656
Optics	4448	Nursing	612
Physics, Atomic, Molecular & Chemical	6051	Women's Studies	167
Biophysics	4618	Area Studies	511
Chemistry, Organic	6518	Political Science	1377
Chemistry, Analytical	5024	Criminology & Penology	309
Medical Laboratory Technology	864	International Relations	831
Chemistry, Physical	10261	Education, Special	127
Materials Science, Composites	942	Archeology	530
Reproductive Biology	1894	History	4037
Genetics & Heredity	6511	Family Studies	210
Immunology	8295	History Of Social Sciences	519

Table 7.15. (Cont.)

Category	Output	Category	Output
Engineering, Electrical & Electronic	7585	Psychology, Psychoanalysis	126
Chemistry, Inorganic & Nuclear	4819	Language & Linguistics	986
Physics, Condensed Matter	10434	Psychology, Applied	483
Automation & Control Systems	1347	Health Policy & Services	378
Meteorology & Atmospheric Sciences	3437	Philosophy	990
Behavioral Sciences	1598	Ethnic Studies	155
Toxicology	2647	Industrial Relations & Labor	182
Geriatrics & Gerontology	1055	Communication	323
Engineering	1266	Demography	209
Agricultural Economics & Policy	119	Engineering, Geological	344
Forestry	1007	Health Care Sciences & Services	1329
History & Philosophy Of Science	594	Literature, Romance	753
Computer Science, Artificial Intelligence	3325	Theater	120
Engineering, Manufacturing	903	Literary Reviews	180
Infectious Diseases	4009	Literature, Slavic	10
Rheumatology	2137	Engineering, Ocean	249
Urology & Nephrology	3832	Music	297
Telecommunications	1150	Literature	1519
Paleontology	896	Poetry	60
Allergy	1278	Literature, American	14
Biology, Miscellaneous	905	Arts & Humanities, General	1949
Materials Science, Ceramics	1553	Architecture	100
Public, Environmental & Occupational Health	4431	Art	551
Emergency Medicine & Critical Care	376	Religion	1216

Now, the scientogram below (Fig. 7.11) is the PFNET visualization of the relations of the 316,878 scientific documents produced by the US in the year 2002, as contained in the ISI databases, and grouped according to the 218 JCR categories. The exact correspondence between each category and its production can be viewed in Table 7.16

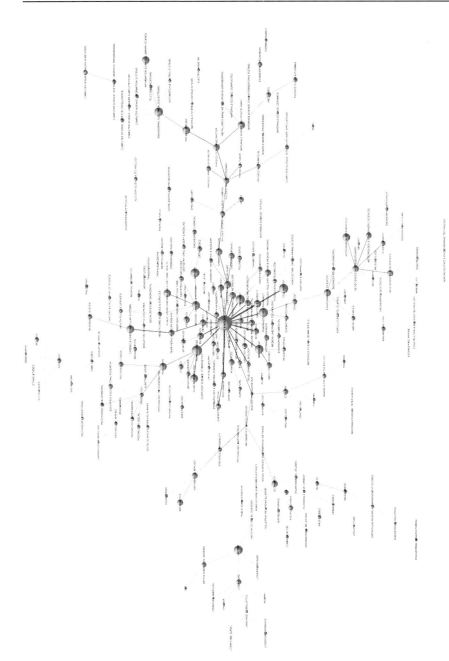

Fig. 7.11. Scientogram of the domain US scientific output, 2002

Table 7.16. Listing of the JCR categories and the US output for 2002

Category	Output	Category	Output
Energy & Fuels	943	Medicine, Legal	357
Geosciences, Interdisciplinary	4330	Anthropology	1341
Engineering, Petroleum	402	Peripheral Vascular Disease	3772
Gastroenterology & Hepatology	2689	Rehabilitation	1553
Radiology, Nuclear Medicine & Medical Imaging	4139	Sport Sciences	2260
Mathematics	3637	Instruments & Instrumentation	1747
Medicine, General & Internal	6733	Geochemistry & Geophysics	3891
Education, Scientific Disciplines	835	Mineralogy	420
Medical Informatics	515	Engineering, Environmental	1381
Entomology	1612	Computer Science, Hardware & Architecture	1207
Chemistry, Multidisciplinary	4887	Developmental Biology	1828
Construction & Building Technology	554	Andrology	101
Materials Science, Multidisciplinary	5236	Engineering, Biomedical	1462
Computer Science, Theory & Methods	2262	Oceanography	2503
Computer Science, Information Systems	1844	Psychology	1749
Computer Science, Software, Graphics, Programming	1698	Nuclear Science & Technology	1442
Mathematics, Applied	3601	Chemistry, Applied	887
Acoustics	975	Physics, Fluids & Plasmas	2033
Cardiac & Cardiovascular Systems	5763	Information Science & Library Science	5556
Respiratory System	2327	Physics, Particles & Fields	2413
Agriculture, Dairy & Animal Science	1156	Chemistry, Medicinal	1818
Agriculture	1382	Materials Science, Paper & Wood	213
Agriculture, Soil Science	694	Ergonomics	209
Nutrition & Dietetics	2194	Spectroscopy	1430
Food Science & Technology	1840	Operations Research & Management Science	1302
Anesthesiology	1375	Engineering, Marine	31
Anatomy & Morphology	402	Thermodynamics	806
Engineering, Aerospace	1060	Materials Science, Coatings & Films	642

Table 7.16. (Cont.)

Category	Output	Category	Output
Astronomy & Astrophysics	5800	Fisheries	938
Biochemistry & Molecular Biology	21694	Limnology	560
Plant Sciences	3459	Mathematics, Miscellaneous	482
Zoology	2493	Geography	642
Biology	1833	Ornithology	361
Biotechnology & Applied Microbiology	3934	Materials Science, Biomaterials	408
Surgery	8154	Geology	491
Crystallography	841	Computer Science, Interdisciplinary Applications	1923
Biochemical Research Methods	3127	Horticulture	745
Cell Biology	8440	Computer Science, Cybernetics	221
Dermatology & Venereal Diseases	1628	Transplantation	1482
Endocrinology & Metabolism	4410	Psychology, Experimental	1462
Hematology	4138	Psychology, Mathematical	250
Environmental Sciences	4900	Materials Science, Characterization & Testing	267
Water Resources	1604	Engineering, Civil	1826
Marine & Freshwater Biology	1879	Mining & Mineral Processing	327
Metallurgy & Metallurgical Engineering	1092	Physics, Mathematical	1964
Mechanics	2443	Electrochemistry	691
Medicine, Research & Experimental	3849	Psychology, Developmental	1574
Neurosciences	10662	Education & Educational Research	2339
Clinical Neurology	6257	Engineering, Industrial	775
Pathology	2399	Mycology	257
Obstetrics & Gynecology	2582	Psychology, Educational	761
Dentistry, Oral Surgery & Medicine	1496	Environmental Studies	1122
Ecology	3043	Social Sciences, Biomedical	399
Oncology	8338	Psychology, Biological	612
Ophthalmology	2263	Sociology	2198
Orthopedics	2346	Law	2146
Otorhinolaryngology	1436	Microscopy	228
Pediatrics	4073	Remote Sensing	441
Pharmacology & Pharmacy	8341	Transportation	134
Physics, Multidisciplinary	3560	Psychology, Clinical	2634

Table 7.16. (Cont.)

Category	Output	Category	Output
Physiology	4242	Imaging Science & Photographic Technology	367
Polymer Science	1949	Materials Science, Textiles	177
Physics, Applied	6006	Economics	4193
Engineering, Chemical	2182	Social Sciences, Mathematical Methods	565
Psychiatry	5565	Business	1603
Tropical Medicine	243	Management	1512
Parasitology	577	Urban Studies	739
Veterinary Sciences	3094	Social Sciences, Interdisciplinary	1112
Virology	2049	Planning & Development	786
Substance Abuse	1061	Business, Finance	1017
Engineering, Mechanical	2371	Social Sigues	433
Microbiology	4059	Public Administration	464
Statistics & Probability	1868	Social Work	688
Physics, Nuclear	1534	Psychology, Social	1255
Optics	3008	Nursing	1288
Physics, Atomic, Molecular & Chemical	2986	Women's Studies	583
Biophysics	3440	Area Studies	1346
Chemistry, Organic	3635	Political Science	2386
Chemistry, Analytical	2950	Criminology & Penology	445
Medical Laboratory Technology	1014	International Relations	734
Chemistry, Physical	4923	Education, Special	459
Materials Science, Composites	454	Archeology	629
Reproductive Biology	1176	History	7807
Genetics & Heredity	5318	Family Studies	956
Immunology	7379	History Of Social Sciences	585
Engineering, Electrical & Electronic	7509	Psychology, Psychoanalysis	253
Chemistry, Inorganic & Nuclear	1750	Language & Linguistics	725
Physics, Condensed Matter	4290	Psychology, Applied	1220
Automation & Control Systems	877	Health Policy & Services	1624
Meteorology & Atmospheric Sciences	3478	Philosophy	1255
Behavioral Sciences	1884	Ethnic Studies	127
Toxicology	2784	Industrial Relations & Labor	462
Geriatrics & Gerontology	1207	Communication	972
Engineering	1096	Demography	327
Agricultural Economics & Policy	237	Engineering, Geological	342

Table 7.16. (Cont.)

Category	Output	Category	Output
Forestry	819	Health Care Sciences & Services	2136
History & Philosophy Of Science	974	Literature, Romance	1059
Computer Science, Artificial Intelligence	1654	Theater	495
Engineering, Manufacturing	786	Literary Reviews	456
Infectious Diseases	3239	Literature, Slavic	112
Rheumatology	957	Engineering, Ocean	277
Urology & Nephrology	3064	Music	679
Telecommunications	1299	Literature	2485
Paleontology	348	Poetry	212
Allergy	580	Literature, American	502
Biology, Miscellaneous	891	Arts & Humanities, General	2371
Materials Science, Ceramics	502	Architecture	296
Public, Environmental & Occupational Health	5767	Art	641
Emergency Medicine & Critical Care	803	Religión	1984

A first appraisal would tell us that the two scientograms (Figs. 7.10 and 7.11) are largely similar. Both have the appearance of a human neuron with a large central axon or neurite. But more careful observation reveals differences. For example, the axon of the US scientogram seems larger than that of the EU. Furthermore, the ramifications around each one of these axons are more densely populated in the EU scientogram.

For a more detailed comparison, we can carry out a macrostructural analysis based on the FA of each of these domains, and then descend to a microstructural analysis through factor scientograms, to finish up with a comparison of the backbone of each.

7.2.2 Factor Analysis

The FA identifies 34 factors in the cocitation matrix of 218 x 218 categories of the EU. Of these, 15 were extracted in view of the scree test (those that have an eigenvalue greater than or equal to one). These accumulated a total of 68.7% of the variance. To capture the nature and categorize each one of these factors, we followed the same methodology as in Sect. 7.1.2. (Factor Analysis). The result is shown in Table 7.17.

Table 7.17. Factors extracted from the domain EU scientific output, 2002

No.	Factor	Eigenvalue	% variance	% accumulative variance
1	Biomedicine	42.74	19.6	19.6
2	Materials Sciences and Applied Physics	22.93	10.5	30.1
3	Management, Law and Economics	14.95	6.9	37
4	Earth & Spaces Sciences	13.25	6.1	43.1
5	Psychology	10.43	4.8	47.8
6	Computer Science and Telecommunications	7.89	3.6	51.5
7	Animal Biology, Ecology	7.04	3.2	54.7
8	Humanities	5.33	2.4	57.1
9	Nuclear Physics and Particle Physics	5.07	2.3	59.5
10	Health Policy, Medical Services	4.62	2.1	61.6
11	Mechanical Engineering	4.13	1.9	63.5
12	Orthopedics	3.43	1.6	65
13	Applied Mathematics	3	1.4	66.4
14	Chemistry	2.66	1.2	67.6
15	Agriculture and Soil Sciences	2.29	1.1	68.7

The tables with the categories that make up each one of the 15 factors of the EU, along with their corresponding factor loadings, are given in Annex III.

In the case of the US domain, the factors identified with an eigenvalue of one or more from the original cocitation matrix of 218 x 218 productive categories were 29. With the scree test, we extracted 14 of these, which again coincide with the ones having an eigenvalue greater than or equal to one. These 14 factors accumulate 69.8% of the variance. The method used to categorize the extracted factors was the same as used for the EU domain.

The tables with the categories that make up each of the 14 US factors, along with the corresponding factor loadings, are included in Annex III.

The prime difference is evident and stems from the number of factors that make up each domain. The domain of the EU has 15 factors as opposed to the 14 of the US. The difference is the factor *Chemistry*, as deduced from the Tables 7.17 and 7.18 (and the factor scientograms of each domain). The reason for this difference is that the categories that make up the factor Chemistry in the EU domain appear identified in the US domain as members of other factors: *Chemistry Inorganic & Nuclear, Engineering Chemical, Crystallography, Chemistry Medical* and *Chemistry Multidisciplinary*, in the factor denominated *Materials Sciences and Applied Physics*. It may also happen that they are not detected by any other factor: *Chemistry Organic, Chemistry Applied* and *Material Sciences Textiles*. However, despite the lesser number of factors, the amount of accumulated variance for the US domain is somewhat greater than that of the EU domain:

Table 7.18. Factors extracted from the domain US scientific output, 2002

No.	Factor	Eigenvalue	% variance	% of accumulative variance
1	Biomedicine	44.632	20.5	20.5
2	Psychology	25.533	11.7	32.2
3	Materials Sciences and Applied Physics	15.928	7.3	39.5
4	Earth & Spaces Sciences	13.237	6.1	45.6
5	Management, Law and Economics	10.279	4.7	50.3
6	Computer Science and Telecommunications	8.282	3.8	54.1
7	Animal Biology, Ecology	6.517	3	57.1
8	Humanities	6.055	2.8	59.8
9	Health Policy, Medical Services	4.463	2	61.9
10	Agriculture and Soil Sciences	4.282	2	63.9
11	Mechanical Engineering	4.099	1.9	65.7
12	Orthopedics	3.411	1.6	67.3
13	Applied Mathematics	3.029	1.4	68.7
14	Nuclear Physics and Particle Physics	2.434	1.1	69.8

69.8% versus 68.7%, respectively. This means that the 14 factors detected in the case of the US explain more about the totality of the domain than the 15 of the EU. It also indicates that the relations between the categories of the US domain factors are stronger and more concentrated than in the EU. Look, for instance, at factor number six of the two domains: *Computer Science and Telecommunication*, where US has 3.8% of variance with eight categories reaching a factor loading of 0.5 or more, whereas the EU accumulates 3.6%, with ten qualifying categories. Also interesting is that the category *Information Science and Library Science* is identified as a member of this factor in the EU domain, but in the US it is not factorized. It may be that the research in the LIS domain within Europe is more focused on Computer Science, whereas US studies are more heavily focused on Library Science in particular.

The second difference resides in the order of the factors of each domain and therefore in the amount of variance accumulated by each factor. True, that the number of categories constituting each factor is conditioned by the journals accepted by the JCR and by the categories it assigns them, provoking factors integrated by a high number of categories like *Biomedicine,* along with ones with very few factors, like *Orthopedics,* or *Agriculture and Soil Sciences*; this in fact occurs in all domains to the same extent. So if one domain differs from another in the order of its factors, it is an indication of the discrepancy that exists insofar as the variance accumulated by each factor in each one of the domains studied. For instance, if we compare the order of the factors of Tables 7.17 and 7.18, we see that *Psychology* occupies the second place in the US domain, with 11.7% of variance, while in the EU it is in fifth place with 4.8% of variance. This tells us that the categories that make up the thematic area of *Psychology* in the US are much more closely related with those conforming the same area in the EU. We could say the same of *Materials Sciences and Applied Physics*, of *Management, Law and Economics*, and of *Nuclear and Particle Physics* in the EU domain, as compared to the US. This evidences the value that industry and energy wield on the European continent, and the difference in attention ceded to the factor *Management, Law and Economics* in Europe with respect to the US.

Factor Scientograms

Like the world factor scientogram, and with the intention of making recognizable at first glance each of the thematic areas as well as the categories within them, we assigned the same color to categories constituting one same factor. Those that are not identified, and do not belong to any thematic area, are colored in dark gray.

The lower left section of each scientogram holds the legend establishing equivalence of color and area, and the name given to each. To aid visual association between the size of the category and its actual output, a sphere of reference (size equivalent to one thousand documents) is given. Finally, all categories belonging to more than one thematic area are shown in red.

Of the 218 categories that constitute the factor scientograms of the domains EU and US, the one to show more factorized categories is the EU, with 198 (20 go unfactorized) (Fig. 7.12); whereas the US domain has 192 factorized and 26 unfactorized categories (Fig. 7.13). There are 11 categories left unfactorized by either domain, listed in Table 7.19:

Incidentally, none of these categories was factorized in the factor scientogram of the world, which can be seen as a logical consequence of the fact that the two domains compared here produce 65.1% of scientific documents worldwide, and so the scientific conduct of both will necessarily be reflected in world output.

The categories not factorized by the EU domain (yet accounted for in the US) are shown in the Table 7.20, along with the thematic area where they were indeed factorized by the US domain.

Table 7.19. Categories not factorized by either of the two domains

Category
Materials Science, Biomaterials
Music
History & Philosophy Of Science
Philosophy
Engineering, Manufacturing
Engineering, Environmental
Materials Science, Paper & Word
Engineering, Industrial
Mathematics
Archeology
Transportation

Table 7.20. Categories not factorized by EU, but factorized by US

Category	Thematic area
Food Science & Technology	Biomedicine
Rehabilitation	Psychology
Art	Humanities
Engineering, Marine	Earth & Spaces Sciences
Theater	Humanities
Poetry	Humanities
Architecture	Management, Law and Economics
Literature, American	Humanities
Imaging Science & Photographic Technology	Earth & Spaces Sciences

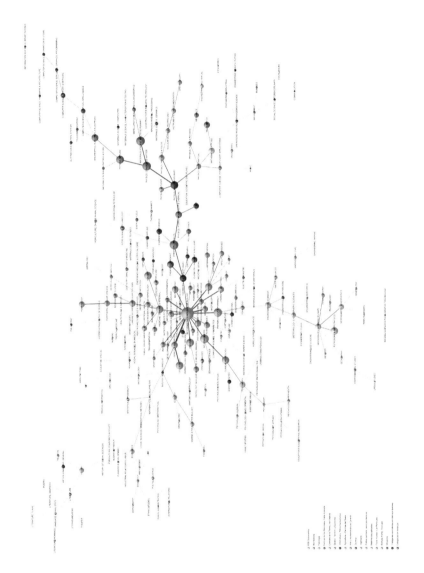

Fig. 7.12. Factor scientogram of the EU scientific domain, 2002

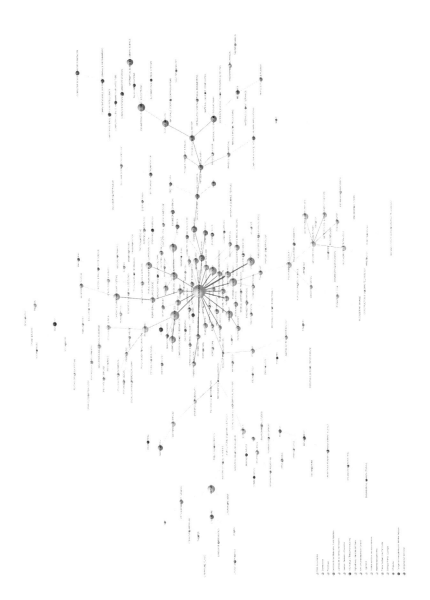

Fig. 7.13. Factor scientogram of the US scientific domain, 2002

Similarly, the categories not factorized by the US domain but indeed factorized by the EU domain in its thematic areas are the 15 indicated in Table 7.21.

7.2.3 Domain Comparison

Macrostructure

The two domains present the same sort of central–peripheral structure, with a major thematic area serving as the point of interconnection of other smaller subject areas that surround them. Likewise, the two domains present the typical bibliometric distribution: just a few areas of great size, and many small ones.

As happens in the factor scientogram of the world, in both the EU and the US we detect what we referred to previously as the typical macrostructural vertebration scheme of the science in developed countries, with the biomedical and earth sciences in the center, the hard sciences to the right, and the soft sciences toward the left.

Table 7.21. Categories included in EU but not factorized by the US domain

Category	Thematic area
Medicine, Legal	Biomedicine
Behavioral Sciences	Etiology
Chemistry, Organic	Chemistry
Chemistry, Applied	Chemistry
Computer Science, Interdisciplinary Applications	Nuclear Physics and Particle Physics
Operations Research & Management Science	Computer Science and Telecommunications
Materials Science, Textiles	Chemistry
Management	Management, Law and Economics
Energy & Fuels	Materials Sciences and Applied Physics
Business	Management, Law and Economics
Information Science & Library Science	Computer Science and Telecommunications
Literature, Slavic	Humanities
Metallurgy & Metallurgical Engineering	Materials Sciences and Applied Physics
Literature, Romance	Humanities
Literature	Humanities

From a relative standpoint, the positions that the factors occupy in their different scientograms are basically the same. There is only one appreciable difference in the visual sense, and that is the situation of *Psychology*. This difference owes more to criteria of spatial ordering of the representation algorithm itself than to structural variations. Apparent structural differences actually point to distinctive modes of connection of the various thematic areas, that is in the sequences or paths of linked categories. More on this is discussed below.

Centrality

If we were to build a network of thematic areas for each one of the factor scientograms and link up these areas by their paths of connection, we would conclude that the most central thematic area of these two domains is *Biomedicine*, as shown in Table 7.22. In both the US and the EU it is the most universal area, and therefore the one contributing most to the scientific advancement of its respective domain.

Table 7.22. Centrality degree of the thematic areas in the domains of EU and US.

EU domain		USA domain	
Thematic area	Grade	Thematic area	Grade
Biomedicine	7	Biomedicine	7
Materials Sciences and Applied Physics	4	Materials Sciences and Applied Physics	4
Animal Biology, Ecology	2	Animal Biology, Ecology	2
Management, Law and Economics	2	Management, Law and Economics	2
Applied Mathematics	2	Applied Mathematics	2
Chemistry	2	Agriculture and Soil Sciences	1
Agriculture and Soil Sciences	1	Earth & Spaces Sciences	1
Earth & Spaces Sciences	1	Nuclear Physics and Particle Physics	1
Nuclear Physics and Particle Physics	1	Humanities	1
Humanities	1	Computer Science and Telecommunications	1
Computer Science and Telecommunications	1	Mechanical Engineering	1
Mechanical Engineering	1	Orthopedics	1
Orthopedics	1	Health Policy, Medical Services	1
Health Policy, Medical Services	1	Psychology	1
Psychology	1		

A quick comparison of the degree of centrality shows very little contrast between the two domains. In fact, they would be identical if not for the absence of the thematic area of *Chemistry* within the US domain.

The degree of nearness or distance of the rest of the thematic areas with respect to *Biomedicine*, indicative of the degree of universality or exclusivity of sources in these areas, is the same for the two domains. Both coincide in that *Computer Science and Telecommunications*, and *Humanities* are the areas that exchange the least information with their neighbors.

Prominence

Using the robbery algorithm as we did with categories, both domains identify *Biomedicine* as the most prominent thematic area, in view of the nature and type of relations maintained with the rest of the areas. Yet the coincidences stop here, as the following tables (Table 7.23) attest.

Whereas *Biomedicine* is the most prominent category in both domains, it is so to a lesser extent in the EU. It practically configures the center of research in the US domain, where it is only minimally accompanied by the area of *Management, Law and Economics*, and *Applied Mathematics*. The European case is more balanced, and while research in the area of *Biomedicine* predominates, it is split up and spread over areas such as *Materials Sciences and Applied Physics*; *Management, Law and Economics*; and *Applied Mathematics*.

Paths of Interconnection Between Thematic Areas

Resorting once again to the factor scientograms of domains EU and US domains, the first difference we encounter is in how *Biomedicine* and *Psychology* are connected. In the EU domain, their path goes *Biochemistry & Molecular Biology* ←→ *Neurosciences* ←→ *Clinical Neurology* ←→ *Psychiatry* ←→ *Psychology*. This leads us to assume that in the EU,

Table 7.23. Prominence of thematic areas in the domains EU and US

EU domain		USA domain	
Thematic area	Prominence	Thematic area	Prominence
Biomedicine	15.27	Biomedicine	21
Materials Sciences and Applied Physics	7.72	Management, Law and Economics	3
Management, Law and Economics	3	Applied Mathematics	2
Applied Mathematics	2		

research in *Psychology* is focused more on clinical and pathological studies–hence its intermediate connection with *Clinical Neurology*–whereas in the US, the material is more theoretical and dedicated to the individual psyche.

Again, one of the most important differences is the absence of the thematic area of *Chemistry* in the US domain. This causes *Biomedicine* to link directly with *Materials Sciences and Applied Physics*; yet in Europe, it is precisely *Chemistry* which is the bridge between those two. This means that, although the path of connection between *Biomedicine* and *Materials Sciences and Applied Physics* is the same for the two domains, the sources related with the categories integrated by the area of *Chemistry* are more profusely utilized by European researchers than by their American counterparts.

Whereas in the US *Mechanic Engineering* is closely related with *Materials Sciences and Applied Physics*, in the EU it is more diffuse amid these areas and *Nuclear and Particle Physics*.

In the EU domain, the area of *Animal Biology and Ecology* has a greater union with studies related with *Genetics and Heredity*. Thus, the path that joins it with *Biomedicine* proceeds in this fashion: *Biology Miscellaneous* ←→ *Genetics and Heredity* ←→ *Biochemistry & Molecular Biology*. In the US, however, research carried out in this area is focused mainly on *Biology* in a strict sense: *Biology Miscellaneous* ←→ *Biology* ←→ *Biochemistry & Molecular Biology*.

Meanwhile, *Applied Mathematics* exhibits the same path of connection in the two domains, and we would like to underline its significance: *Biology Miscellaneous* ←→ *Mathematics Miscellaneous* ←→ *Social Sciences Mathematics*; in both cases it is the bridge connecting *Management, Law and Economics*, and *Humanities*, with all the other areas.

Points of Interaction Among Thematic Areas

A series of categories appear in red in the scientograms. They indicate the zones of friction or points of interaction of the different thematic areas. The comparative study of domains, when centered on these categories responsible for the interchange of information among areas, reveal similarities and differences in the lines of research nourished by intellectual interchanges in the two domains of study.

In both the EU and the US, the level of interaction among thematic area is quite low. At the very most one category is seen belonging to two separate thematic areas.

The multidisciplinary categories coinciding in the domain of the EU and that of the US with their corresponding thematic areas are listed in Table 7.24.

Table 7.24. Categories with double thematic ascription in the domains EU and US

Category	Thematic areas	
Business, Finance	Management, Law and Economics	Applied Mathematics
Communication	Management, Law and Economics	Psychology
Crystallography	Chemistry	Materials Sciences and Applied Physics
Materials Science, Composites	Mechanical Engineering	Materials Sciences and Applied Physics
Otorhinolaryngology	Biomedicine	Orthopedics
Psychology, Mathematical	Psychology	Applied Mathematics
Zoology	Biomedicine	Animal Biology, Ecology

The interactions among thematic areas taking place in the domain of the EU, but not in the US, are given in Table 7.25.

Finally, the similarities among thematic areas appearing in the US but not in the EU are given in Table 7.26.

Table 7.25. Categories with double thematic ascription in the EU domain

Category	Thematic areas	
Anesthesiology	Biomedicine	Orthopedics
Arts & Humanities, General	Humanities	Management, Law and Economics
Biology, Miscellaneous	Biomedicine	Animal Biology, Ecology
Chemistry, Medicinal	Biomedicine	Chemistry
Chemistry, Multidisciplinary	Chemistry	Materials Sciences and Applied Physics
Engineering, Chemical	Chemistry	Materials Sciences and Applied Physics
Optics	Nuclear Physics and Particle Physics	Materials Sciences and Applied Physics
Physics, Multidisciplinary	Nuclear Physics and Particle Physics	Materials Sciences and Applied Physics
Social Sciences, Interdisciplinary	Management, Law and Economics	Psychology
Social Work	Health Policy, Medical Services	Psychology
Women's Studies	Health Policy, Medical Services	Psychology

Table 7.26. Categories with double thematic ascription in the US domain

Category	Thematic area	
Emergency Medicine & Critical Care	Biomedicine	Health Policy, Medical Services
Engineering	Mechanical Engineering	Materials Sciences and Applied Physics
Entomology	Animal Biology, Ecology	Biomedicine
Forestry	Agriculture and Soil Sciences	Animal Biology, Ecology
Materials Science Characterization Test	Mechanical Engineering	Materials Sciences and Applied Physics
Social Studies	Psychology	Management, Law and Economics
Social Sciences, Biomedical	Health Policy, Medical Services	Psychology
Sociology	Sociology	Management, Law and Economics

Microstructure

The EU and the US domain share the same 218 categories. Accordingly, they dispose of 217 links to put these categories together, leaving none isolated. The scientograms of both realms offer a clear example of the hyperbolic nature of bibliometric distribution, demonstrating a greater concentration of larger categories in the center and right-center of the scientogram, and less to the left. We see, then, a greater overall production of documents by the categories pertaining to the medical and "hard" sciences than to those associated with the softer sciences. Likewise, both domains follow the now familiar pattern of central–peripheral connection to lend cohesion to the scientogram structure.

Centrality

Very clearly, in both scientograms (Figs. 7.14 and 7.15), the most central category is *Biochemistry & Molecular Biology*, yet we shall nonetheless use the measures of centrality to determine its degree in each of the domains (Table 7.27).

Biochemistry & Molecular Biology is the most central (and the most universal) of all the categories of either domain, as it is the one with the most common sources. We could say it is the protagonist of scientific advancement in the respective domains, but to a somewhat greater extent in the US, as demonstrated by its greater nodal degree.

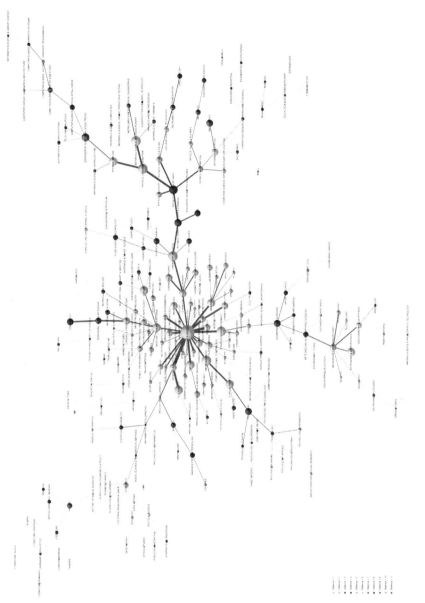

Fig. 7.14. Scientogram of the distances of the domain EU, with respect to its central category

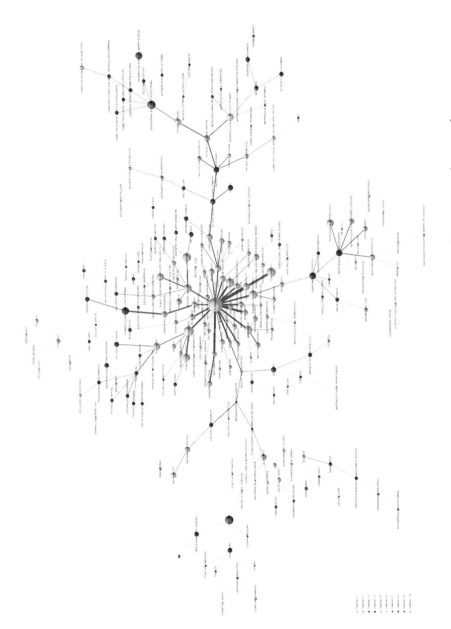

Fig. 7.15. Scientogram of the distances of the US scientific domain with respect to its central category

Table 7.27. Centrality of degree in the top 15 categories of EU and US domains

Eu Domain Category	Grade	Us Domain Category	Grade
Biochemistry & Molecular Biology	26	Biochemistry & Molecular Biology	31
Medicine, General & Internal	10	Economics	11
Economics	10	Geosciences, Interdisciplinary	9
Geosciences, Interdisciplinary	8	Medicine, General & Internal	9
Materials Science, Multidisciplinary	7	Psychology	8
Psychiatry	7	Materials Science, Multidisciplinary	7
Public, Environmental & Occupational Health	7	Public, Environmental & Occupational Health	7
Physics, Multidisciplinary	6	Chemistry, Multidisciplinary	6
Immunology	6	Environmental Sciences	6
Literature	6	Neurosciences	6
Chemistry, Multidisciplinary	5	Engineering, Electrical & Electronic	6
Surgery	5	Physics, Multidisciplinary	5
Environmental Sciences	5	Psychiatry	5
Neurosciences	5	Immunology	5
Chemistry, Physical	5	Sociology	5

The grade of universality and participation of the rest of the categories in the development of each domain will depend on the geodesic distance of each with regard to its respective central category (Tables 7.28 and 7.29).

Table 7.28. Distances of the categories of the scientogram of the EU scientific domain, with regards to its central category

Category	Distance	Category	Distance
Biochemistry & Molecular Biology	0	Limnology	4
Medicine, General & Internal	1	Ornithology	4
Entomology	1	Psychology, Mathematical	4
Chemistry, Multidisciplinary	1	Psychology, Developmental	4
Nutrition & Dietetics	1	Psychology, Clinical	4
Plant Sciences	1	Materials Science, Textiles	4
Zoology	1	Social Sciences, Mathematical Methods	4
Biology	1	Criminology & Penology	4
Biotechnology & Applied Microbiology	1	Education, Special	4

Table 7.28. (Cont.)

Category	Distance	Category	Distance
Biochemical Research Methods	1	Archeology	4
Dermatology & Venereal Diseases	1	Psychology, Psychoanalysis	4
Endocrinology & Metabolism	1	Philosophy	4
Medicine, Research & Experimental	1	Geosciences, Interdisciplinary	5
Neurosciences	1	Engineering, Petroleum	5
Pathology	1	Education, Scientific Disciplines	5
Oncology	1	Medical Informatics	5
Pharmacology & Pharmacy	1	Agriculture, Dairy & Animal Science	5
Physiology	1	Engineering, Aerospace	5
Parasitology	1	Physics, Nuclear	5
Microbiology	1	Physics, Condensed Matter	5
Biophysics	1	Physics, Particles & Fields	5
Medical Laboratory Technology	1	Fisheries	5
Genetics & Heredity	1	Psychology, Experimental	5
Immunology	1	Engineering, Civil	5
Geriatrics & Gerontology	1	Physics, Mathematical	5
Developmental Biology	1	Psychology, Educational	5
Microscopy	1	Economics	5
Gastroenterology & Hepatology	2	Psychology, Social	5
Radiology, Nuclear Medicine & Medical Imaging	2	Family Studies	5
Cardiac & Cardiovascular Systems	2	Psychology, Applied	5
Agriculture	2	Engineering, Ocean	5
Surgery	2	Architecture	5
Crystallography	2	Art	5
Cell Biology	2	Materials Science, Multidisciplinary	6
Clinical Neurology	2	Mathematics, Applied	6
Obstetrics & Gynecology	2	Astronomy & Astrophysics	6
Ophthalmology	2	Physics, Applied	6
Pediatrics	2	Agricultural Economics & Policy	6
Veterinary Sciences	2	Paleontology	6
Virology	2	Geochemistry & Geophysics	6
Chemistry, Organic	2	Oceanography	6
Chemistry, Analytical	2	Nuclear Science & Technology	6

Table 7.28. (Cont.)

Category	Distance	Category	Distance
Chemistry, Physical	2	Physics, Fluids & Plasmas	6
Chemistry, Inorganic & Nuclear	2	Engineering, Marine	6
Behavioral Sciences	2	Geography	6
Toxicology	2	Geology	6
Forestry	2	Computer Science, Interdisciplinary Applications	6
Infectious Diseases	2	Education & Educational Research	6
Rheumatology	2	Sociology	6
Urology & Nephrology	2	Law	6
Allergy	2	Planning & Development	6
Biology, Miscellaneous	2	Business, Finance	6
Public, Environmental & Occupational Health	2	Area Studies	6
Medicine, Legal	2	Political Science	6
Sport Sciences	2	History Of Social Sciences	6
Engineering, Biomedical	2	Industrial Relations & Labor	6
Chemistry, Medicinal	2	Engineering, Geological	6
Horticulture	2	Mathematics	7
Computer Science, Cybernetics	2	Construction & Building Technology	7
Transplantation	2	Metallurgy & Metallurgical Engineering	7
Mycology	2	Mechanics	7
Health Care Sciences & Services	2	Materials Science, Composites	7
Acoustics	3	Engineering, Electrical & Electronic	7
Respiratory System	3	Engineering	7
Agriculture, Soil Science	3	Materials Science, Ceramics	7
Anesthesiology	3	Instruments & Instrumentation	7
Anatomy & Morphology	3	Mineralogy	7
Environmental Sciences	3	Operations Research & Management Science	7
Dentistry, Oral Surgery & Medicine	3	Materials Science, Coatings & Films	7
Ecology	3	Materials Science, Characterization & Testing	7
Orthopedics	3	Mining & Mineral Processing	7
Otorhinolaryngology	3	Environmental Studies	7
Polymer Science	3	Remote Sensing	7
Engineering, Chemical	3	Public Administration	7

Table 7.28. (Cont.)

Category	Distance	Category	Distance
Psychiatry	3	International Relations	7
Tropical Medicine	3	History	7
Physics, Atomic, Molecular & Chemical	3	Ethnic Studies	7
Reproductive Biology	3	Demography	7
History & Philosophy Of Science	3	Music	7
Emergency Medicine & Critical Care	3	Engineering, Mechanical	8
Anthropology	3	Automation & Control Systems	8
Peripheral Vascular Disease	3	Computer Science, Artificial Intelligence	8
Rehabilitation	3	Telecommunications	8
Chemistry, Applied	3	Spectroscopy	8
Materials Science, Paper & Wood	3	Engineering, Industrial	8
Mathematics, Miscellaneous	3	Imaging Science & Photographic Technology	8
Materials Science, Biomaterials	3	Management	8
Electrochemistry	3	Urban Studies	8
Social Sciences, Biomedical	3	Literature	8
Psychology, Biological	3	Arts & Humanities, General	8
Transportation	3	Religion	8
Social Work	3	Computer Science, Theory & Methods	9
Nursing	3	Engineering, Manufacturing	9
Women's Studies	3	Ergonomics	9
Health Policy & Services	3	Business	9
Energy & Fuels	4	Social Sciences, Interdisciplinary	9
Food Science & Technology	4	Language & Linguistics	9
Hematology	4	Theater	9
Water Resources	4	Literary Reviews	9
Marine & Freshwater Biology	4	Poetry	9
Physics, Multidisciplinary	4	Literature, American	9
Substance Abuse	4	Computer Science, Software, Graphics, Programming	10
Statistics & Probability	4	Computer Science, Hardware & Architecture	10
Optics	4	Social Issues	10
Meteorology & Atmospheric Sciences	4	Communication	10
Engineering, Environmental	4	Literature, Romance	10

Table 7.28. (Cont.)

Category	Distance	Category	Distance
Andrology	4	Literature, Slavic	10
Psychology	4	Computer Science, Information Systems	11
Thermodynamics	4	Information Science & Library Science	12

Table 7.29. Distances of the categories in the scientogram of the US scientific domain, with respect to the central category

Category	Distance	Category	Distance
Biochemistry & Molecular Biology	0	Forestry	4
Gastroenterology & Hepatology	1	Engineering, Environmental	4
Medicine, General & Internal	1	Ergonomics	4
Entomology	1	Spectroscopy	4
Chemistry, Multidisciplinary	1	Limnology	4
Plant Sciences	1	Ornithology	4
Biology	1	Psychology, Experimental	4
Biotechnology & Applied Microbiology	1	Psychology, Mathematical	4
Crystallography	1	Psychology, Developmental	4
Biochemical Research Methods	1	Education & Educational Research	4
Dermatology & Venereal Diseases	1	Social Sciences, Mathematical Methods	4
Endocrinology & Metabolism	1	Social Sciences, Interdisciplinary	4
Hematology	1	Psychology, Social	4
Medicine, Research & Experimental	1	Education, Special	4
Neurosciences	1	Archeology	4
Pathology	1	Family Studies	4
Oncology	1	Psychology, Applied	4
Ophthalmology	1	Philosophy	4
Pharmacology & Pharmacy	1	Engineering, Petroleum	5
Physiology	1	Mathematics, Applied	5
Parasitology	1	Astronomy & Astrophysics	5
Virology	1	Physics, Nuclear	5
Microbiology	1	Physics, Condensed Matter	5
Biophysics	1	Meteorology & Atmospheric Sciences	5

Table 7.29. (Cont.)

Category	Distance	Category	Distance
Genetics & Heredity	1	Paleontology	5
Immunology	1	Instruments & Instrumentation	5
Geriatrics & Gerontology	1	Geochemistry & Geophysics	5
Urology & Nephrology	1	Oceanography	5
Developmental Biology	1	Physics, Particles & Fields	5
Andrology	1	Materials Science, Paper & Wood	5
Mycology	1	Fisheries	5
Microscopy	1	Geography	5
Radiology, Nuclear Medicine & Medical Imaging	2	Geology	5
Agriculture	2	Engineering, Civil	5
Nutrition & Dietetics	2	Physics, Mathematical	5
Food Science & Technology	2	Psychology, Educational	5
Zoology	2	Sociology	5
Surgery	2	Economics	5
Cell Biology	2	Social Work	5
Clinical Neurology	2	Criminology & Penology	5
Obstetrics & Gynecology	2	Engineering, Geological	5
Pediatrics	2	Mathematics	6
Polymer Science	2	Construction & Building Technology	6
Psychiatry	2	Materials Science, Multidisciplinary	6
Veterinary Sciences	2	Engineering, Aerospace	6
Chemistry, Organic	2	Physics, Applied	6
Chemistry, Analytical	2	Engineering	6
Medical Laboratory Technology	2	Agricultural Economics & Policy	6
Chemistry, Physical	2	Mineralogy	6
Chemistry, Inorganic & Nuclear	2	Nuclear Science & Technology	6
Behavioral Sciences	2	Engineering, Marine	6
Toxicology	2	Computer Science, Interdisciplinary Applications	6
Infectious Diseases	2	Environmental Studies	6
Rheumatology	2	Law	6
Allergy	2	Remote Sensing	6

Table 7.29. (Cont.)

Category	Distance	Category	Distance
Biology, Miscellaneous	2	Business	6
Public, Environmental & Occupational Health	2	Planning & Development	6
Emergency Medicine & Critical Care	2	Business, Finance	6
Medicine, Legal	2	Social Issues	6
Peripheral Vascular Disease	2	Public Administration	6
Sport Sciences	2	Political Science	6
Engineering, Biomedical	2	History Of Social Sciences	6
Chemistry, Applied	2	Ethnic Studies	6
Chemistry, Medicinal	2	Industrial Relations & Labor	6
Horticulture	2	Demography	6
Computer Science, Cybernetics	2	Engineering, Ocean	6
Transplantation	2	Religion	6
Health Care Sciences & Services	2	Metallurgy & Metallurgical Engineering	7
Education, Scientific Disciplines	3	Mechanics	7
Medical Informatics	3	Materials Science, Composites	7
Acoustics	3	Engineering, Electrical & Electronic	7
Cardiac & Cardiovascular Systems	3	Materials Science, Ceramics	7
Agriculture, Dairy & Animal Science	3	Materials Science, Coatings & Films	7
Anesthesiology	3	Materials Science, Characterization & Testing	7
Anatomy & Morphology	3	Mining & Mineral Processing	7
Environmental Sciences	3	Imaging Science & Photographic Technology	7
Dentistry, Oral Surgery & Medicine	3	Management	7
Ecology	3	Urban Studies	7
Orthopedics	3	Area Studies	7
Otorhinolaryngology	3	International Relations	7
Engineering, Chemical	3	History	7
Tropical Medicine	3	Communication	7
Substance Abuse	3	Music	7
Physics, Atomic, Molecular & Chemical	3	Computer Science, Information Systems	8

Table 7.29. (Cont.)

Category	Distance	Category	Distance
Reproductive Biology	3	Engineering, Mechanical	8
History & Philosophy Of Science	3	Automation & Control Systems	8
Anthropology	3	Computer Science, Artificial Intelligence	8
Rehabilitation	3	Telecommunications	8
Psychology	3	Computer Science, Hardware & Architecture	8
Mathematics, Miscellaneous	3	Physics, Fluids & Plasmas	8
Materials Science, Biomaterials	3	Operations Research & Management Science	8
Social Sciences, Biomedical	3	Electrochemistry	8
Psychology, Biological	3	Literary Reviews	8
Transportation	3	Literature	8
Psychology, Clinical	3	Arts & Humanities, General	8
Materials Science, Textiles	3	Architecture	8
Nursing	3	Computer Science, Software, Graphics, Programming	9
Women's Studies	3	Information Science & Library Science	9
Psychology, Psychoanalysis	3	Thermodynamics	9
Health Policy & Services	3	Engineering, Industrial	9
Energy & Fuels	4	Language & Linguistics	9
Geosciences, Interdisciplinary	4	Theater	9
Respiratory System	4	Poetry	9
Agriculture, Soil Science	4	Literature, American	9
Water Resources	4	Art	9
Marine & Freshwater Biology	4	Computer Science, Theory & Methods	10
Physics, Multidisciplinary	4	Engineering, Manufacturing	10
Statistics & Probability	4	Literature, Romance	10
Optics	4	Literature, Slavic	10

As for the geodesic distance of the categories of the two domains with regards to the central category, results are similar. The EU traces a maximum path over 12 nodes, whereas the US trek covers 10. Granted, there is only one category with a maximum distance of 12: precisely, *Information Science & Library Science*.

There are, however, a couple of distinctions that should be pointed out. Table 7.30 gathers the numerical information.

Table 7.30. Differences in geodesic distance between EU and US domains

Category	EU distance	US distance	Difference
Ergonomics	9	4	5
Social Sciences, Interdisciplinary	9	4	5
Spectroscopy	8	4	4
Social Issues	10	6	4
Andrology	4	1	3
Hematology	4	1	3
Business	9	6	3
Communication	10	7	3
Computer Science, Information Systems	11	8	3
Information Science & Library Science	12	9	3

For example, the first two categories on Table 7.30—*Ergonomics* and *Social Sciences Interdisciplinary*—are at a distance of nine links with respect to the central category in the EU domain. In the US domain there is a distance of just four, meaning that in the US these categories participate more actively in research than in Europe. This comes as no surprise, considering their thematic ascription in one domain and the other: *Psychology* in the US, and *Management, Law and Economics* in Europe. We could say the same of *Andrology* and *Hematology*, both of which intervene more in American research than in European studies, as illustrated by the minimal distances from the central category. Here again, the activity of *Information Science & Library Science* is noteworthy: while in the EU the most significant link is with the category *Computer Science Information Systems* as its main object of research, it is also related with the systems of automated information. But in the US, this category is related directly with *Engineering Electrical & Electronics*, the nucleus from which all the research related with *Computers Sciences* stems. This finding leads us to affirm that research is more closely related with Computer Science in general than with any one of its specific disciplines. This same fact would explain the different distances expressed in the two domains.

Prominence

The robbery algorithm selects 29 prominent categories in the domain of the EU and 28 in the US. Just as we saw with centrality, the two domains here identify *Biochemistry & Molecular Biology* as the most prominent area, shown in Table 7.31. However, they only coincide on this point. This category is slightly more prominent in the US than in Europe. The name and order of the categories identified as salient is also different, showing greater prominence in the US in the coincident cases.

Table 7.31. Prominence of the categories in the scientific domains EU and US

EU domain		USA domain	
Category	Prominence	Category	Prominence
Biochemistry & Molecular Biology	156.62	Biochemistry & Molecular Biology	164.97
Economics	28.86	Geosciences, Interdisciplinary	38.67
Psychiatry	27.33	Economics	30.93
Geosciences, Interdisciplinary	25.69	Psychology	23.71
Physics, Multidisciplinary	22.57	Engineering, Electrical & Electronic	23.00
Chemistry, Physical	21.82	Literature	21.53
Literature	21.57	Materials Science, Multidisciplinary	20.75
Materials Science, Multidisciplinary	14.62	Physics, Multidisciplinary	19.25
Biology, Miscellaneous	12.94	Sociology	13.75
Environmental Sciences	12.56	Ecology	11
Engineering, Electrical & Electronic	11.14	Biology, Miscellaneous	7.23
Mathematics, Applied	9	Mathematics, Miscellaneous	6.68
Mathematics, Miscellaneous	9	Mathematics, Applied	5.86
Computer Science, Theory & Methods	6.86	Engineering, Civil	3
Archeology	5.86	Energy & Fuels	3
Engineering, Industrial	5	Radiology, Nuclear Medicine & Medical Imaging	3
Social Sciences, Interdisciplinary	5	Cardiac & Cardiovascular Systems	3
Management	4	Food Science & Technology	3
Radiology, Nuclear Medicine & Medical Imaging	3	Cell Biology	3
Computer Science, Information Systems	3	Rehabilitation	3
Mechanics	3	Instruments & Instrumentation	3
Chemistry, Analytical	3	Engineering, Biomedical	3
Reproductive Biology	3	Computer Science, Interdisciplinary Applications	3
Rehabilitation	3	Engineering, Industrial	3
Instruments & Instrumentation	3	Urban Studies	3
Engineering, Biomedical	3	Peripheral Vascular Disease	2
Engineering, Civil	3	Sport Sciences	2
Environmental Studies	3	Operations Research & Management Science	2
Sport Sciences	2	Management	2
Nuclear Science & Technology	2		

The scree test identifies 13 categories as the most prominent of the EU domain, 12 of which give rise to a thematic area detected by FA. In the case of the US domain, there are also 13 categories identified, and 11 coincide with the origin of a thematic area, as seen in the scientograms (Figs. 7.16 and 7.17) of prominence presented here.

The three categories of each domain that were not detected with the robbery algorithm must be identified by viewing the factor scientogram of each domain and counting the number of links of each category. Interestingly enough, these are the categories *Medicine General & Internal, Plants Science and Surgery*, with the thematic areas to which they respectively give rise: *Health Policy & Medical Services, Earth and Soil Sciences, and Orthopedics*, the same for the two domains. And if we recall, they were also the same ones for the world domain, where we mentioned that they possibly identified specialized fields rather than broader thematic areas.

Backbone

The differences in the vertebration of science from one domain to the next are highly significant. Further scientograms (Figs. 7.18 and 7.19) will illustrate this quite clearly.

The EU aims its investigative efforts basically toward three thematic areas: *Biomedicine, Materials Science & Applied Physics*, and *Chemistry*. Meanwhile, back in the US, focus is essentially on one. These differences are possibly due to distinctive governmental policies regarding research for the representative geographic domain, and above all to the health system that each one relies on: private in the US, and mostly public in the EU.

Whereas in Europe we have 30 categories with a high flux of interchange and informational output, reflected in the thickness of the links and size of the categories, in the US there are "only" 24. Yet *Biomedicine* itself occupies 24 categories in the US, 6 more than the 18 categories of the EU.

7.3 Scientography and the Evolution of Domains

The following analysis is centered on the temporal evolution of the geographic–scientific domain of Spain itself, through a look at three scientograms, each representing 4 years of output, beginning in 1990. The characteristics of these three scientograms are very much the same as the ones we have seen up to now, but taken together they provide sequential information, leading the viewer to infer the evolution of the domain. One peculiarity is that in this case they show weak links we could refer to as dubious, as suggested back in Sect. 6.4.1. These intercategorical connections appear in red.

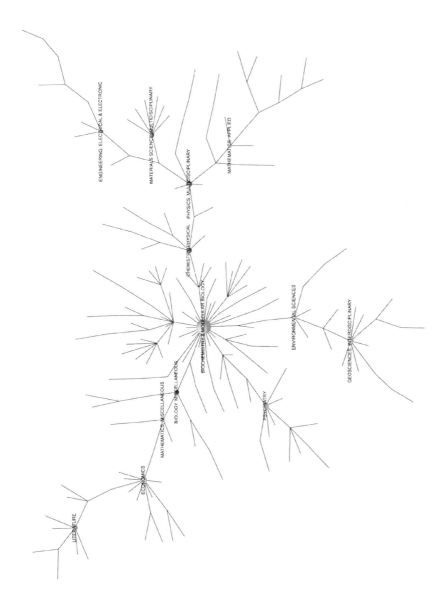

Fig. 7.16. Scientogram of the most prominent categories of the EU scientific domain

Fig. 7.17. Scientogram of the most prominent categories of the US scientific domain

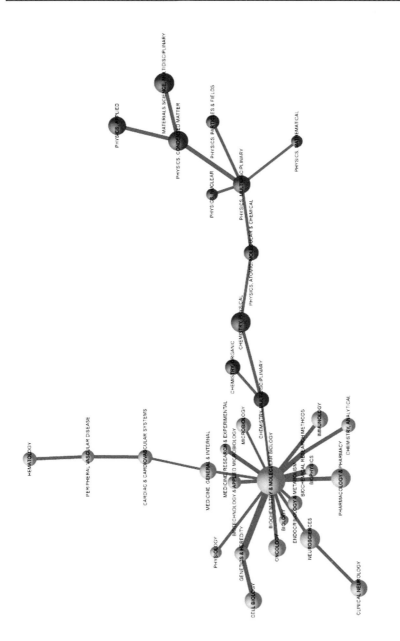

Fig. 7.18. Scientogram of the vertebration of science in the domain EU, 2002

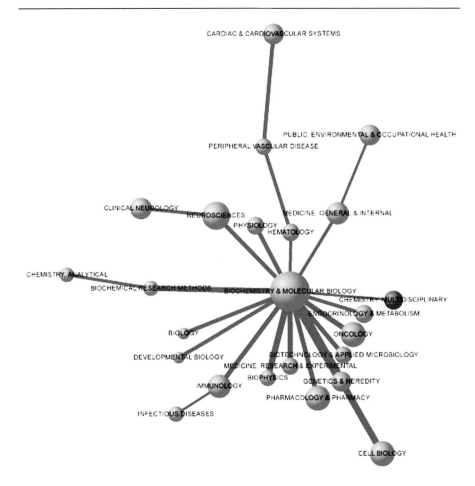

Fig. 7.19. Scientogram of the vertebration of science in the domain US, 2002

7.3.1 Scientograms

Figures 7.20, 7.21, and 7.22 offer the scientograms for Spain's scientific output over the periods 1990–1994, 1995–1998, and 1999–2002. Respectively, they stand as the graphic schematizations of 71,085, 88,166 and 110,286 documents, grouped into 219, 223, and 240 JCR categories. Total production broken down is shown in Tables 7.32, 7.33, and 7.34.

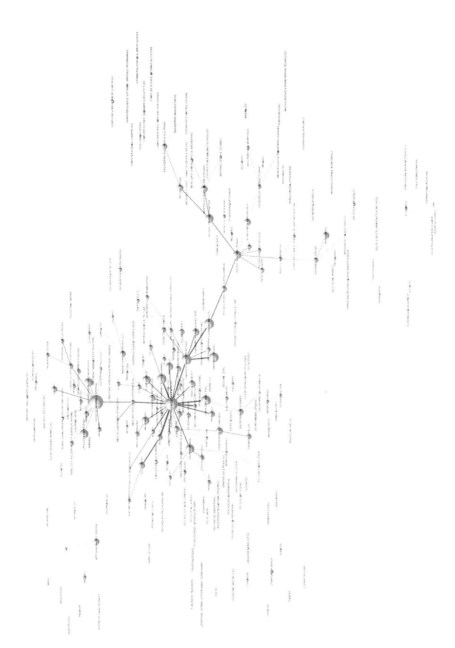

Fig. 7.20. Scientogram of Spain's output 1990–1994

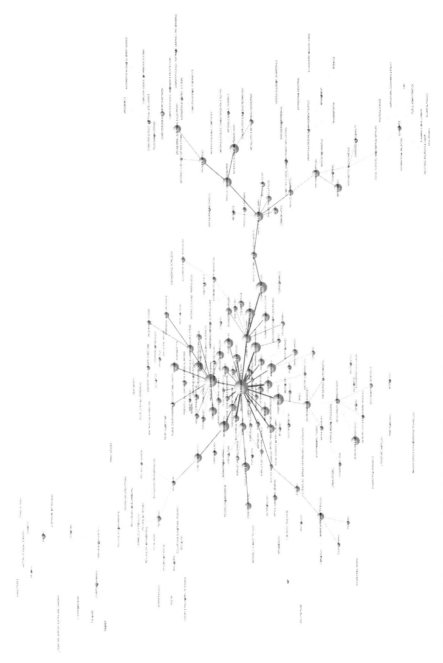

Fig. 7.21. Scientogram of Spain's scientific output for the period 1995–1998

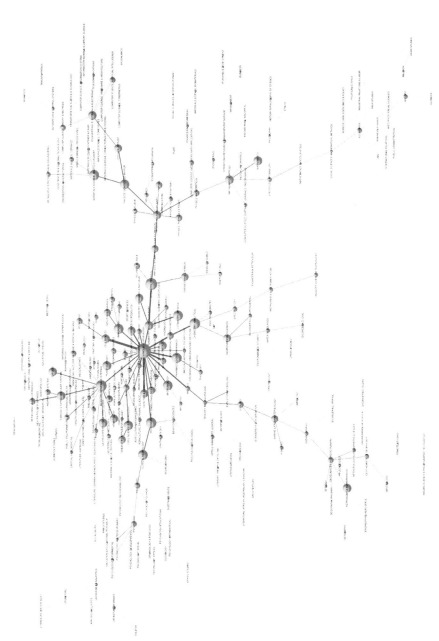

Fig. 7.22. Scientogram of Spain's scientific output for the period 1999–2002

Table 7.32. Listing of JCR categories and Spain's scientific output for the period 1990–1994

Category	Output	Category	Output
Energy & Fuels	345	Anthropology	60
Geosciences, Interdisciplinary	439	Peripheral Vascular Disease	463
Engineering, Petroleum	21	Rehabilitation	19
Gastroenterology & Hepatology	2001	Sport Sciences	48
Radiology, Nuclear Medicine & Medical Imaging	553	Instruments & Instrumentation	456
Mathematics	1306	Geochemistry & Geophysics	292
Medicine, General & Internal	5286	Mineralogy	135
Education, Scientific Disciplines	101	Engineering, Environmental	94
Medical Informatics	52	Computer Science, Hardware & Architecture	151
Entomology	241	Developmental Biology	193
Chemistry, Multidisciplinary	2643	Andrology	27
Construction & Building Technology	65	Engineering, Biomedical	113
Materials Science, Multidisciplinary	1344	Oceanography	180
Computer Science, Theory & Methods	347	Psychology	491
Computer Science, Information Systems	165	Nuclear Science & Technology	414
Computer Science, Software, Graphics, Programming	85	Chemistry, Applied	788
Mathematics, Applied	804	Physics, Fluids & Plasmas	271
Acoustics	70	Information Science & Library Science	115
Cardiac & Cardiovascular Systems	766	Physics, Particles & Fields	446
Respiratory System	443	Chemistry, Medicinal	293
Agriculture, Dairy & Animal Science	294	Materials Science, Paper & Wood	37
Agriculture	972	Ergonomics	5
Agriculture, Soil Science	346	Spectroscopy	498
Nutrition & Dietetics	358	Operations Research & Management Science	84
Food Science & Technology	1363	Engineering, Marine	30
Anesthesiology	103	Thermodynamics	53
Anatomy & Morphology	381	Materials Science, Coatings & Films	139
Engineering, Aerospace	31	Fisheries	179
Astronomy & Astrophysics	1328	Limnology	80
Biochemistry & Molecular Biology	4352	Mathematics, Miscellaneous	80
Plant Sciences	2032	Geography	60

Table 7.32. (Cont.)

Category	Output	Category	Output
Zoology	649	Ornithology	155
Biology	445	Materials Science, Biomaterials	35
Biotechnology & Applied Microbiology	982	Geology	102
Surgery	1064	Computer Science, Interdisciplinary Applications	266
Crystallography	518	Horticulture	135
Biochemical Research Methods	416	Computer Science, Cybernetics	69
Cell Biology	1193	Transplantation	468
Dermatology & Venereal Diseases	670	Psychology, Experimental	153
Endocrinology & Metabolism	1106	Psychology, Mathematical	35
Hematology	902	Materials Science, Characterization & Testing	36
Environmental Sciences	1041	Engineering, Civil	248
Water Resources	296	Mining & Mineral Processing	53
Marine & Freshwater Biology	998	Physics, Mathematical	430
Metallurgy & Metallurgical Engineering	288	Electrochemistry	383
Mechanics	264	Psychology, Developmental	17
Medicine, Research & Experimental	492	Education & Educational Research	60
Neurosciences	2027	Engineering, Industrial	13
Clinical Neurology	732	Mycology	187
Pathology	581	Psychology, Educational	12
Obstetrics & Gynecology	270	Environmental Studies	46
Dentistry, Oral Surgery & Medicine	143	Social Sciences, Biomedical	14
Ecology	522	Psychology, Biological	147
Oncology	877	Sociology	16
Ophthalmology	322	Law	23
Orthopedics	202	Microscopy	90
Otorhinolaryngology	91	Remote Sensing	47
Pediatrics	455	Transportation	8
Pharmacology & Pharmacy	2530	Psychology, Clinical	66
Physics, Multidisciplinary	1818	Imaging Science & Photographic Technology	45
Physiology	1000	Materials Science, Textiles	66
Polymer Science	850	Economics	234
Physics, Applied	1093	Social Sciences, Mathematical Methods	81
Engineering, Chemical	714	Business	36
Psychiatry	371	Management	63
Tropical Medicine	65	Urban Studies	28

Table 7.32. (Cont.)

Category	Output	Category	Output
Parasitology	195	Social Sciences, Interdisciplinary	18
Veterinary Sciences	627	Planning & Development	27
Virology	283	Business, Finance	26
Substance Abuse	112	Social Issues	8
Engineering, Mechanical	122	Public Administration	10
Microbiology	1614	Social Work	4
Statistics & Probability	278	Psychology, Social	62
Physics, Nuclear	618	Nursing	9
Optics	614	Women's Studies	2
Physics, Atomic, Molecular & Chemical	897	Area Studies	10
Biophysics	892	Political Science	37
Chemistry, Organic	2464	Criminology & Penology	4
Chemistry, Analytical	2627	International Relations	23
Medical Laboratory Technology	181	Education, Special	5
Chemistry, Physical	2624	Archeology	28
Materials Science, Composites	16	History	303
Reproductive Systems	189	Family Studies	10
Genetics & Heredity	1332	History Of Social Sciences	20
Immunology	1564	Psychology, Psychoanalysis	5
Engineering, Electrical & Electronic	856	Language & Linguistics	80
Chemistry, Inorganic & Nuclear	1541	Psychology, Applied	32
Physics, Condensed Matter	1913	Health Policy & Services	21
Automation & Control Systems	66	Philosophy	169
Meteorology & Atmospheric Sciences	134	Industrial Relations & Labor	4
Behavioral Sciences	176	Communication	14
Toxicology	594	Demography	17
Geriatrics & Gerontology	91	Literature, Romance	353
Engineering	114	Theater	30
Agricultural Economics & Policy	10	Literary Reviews	7
Forestry	127	Music	7
History & Philosophy Of Science	62	Literature	77
Computer Science, Artificial Intelligence	104	Poetry	1
Engineering, Manufacturing	8	Arts & Humanities, General	971
Infectious Diseases	688	Architecture	7
Rheumatology	431	Art	135
Urology & Nephrology	2141	Religion	46
Telecommunications	126	Folklore	33
Paleontology	97	Film, Radio, Television	5
Allergy	433	Classics	49
Biology, Miscellaneous	131	Asian Studies	10

Table 7.32. (Cont.)

Category	Output	Category	Output
Materials Science, Ceramics	189	Literature, German, Netherlandic, Scandinavian	2
Public, Environmental & Occupational Health	322	Literature, British Isles	9
Emergency Medicine & Critical Care	141	Literature, African, Australian, Canadian	3
Medicine, Legal	66		

The scientogram of this period is made up of 219 JCR categories, two of which appear isolated. These are *Poetry* and *Literature German, Netherlandic, Scandinavian*, which, while having articles in the ISI databases, are not shown to have any relations with categories other than themselves.

Table 7.33. Listing of the JCR categories and Spains's scientific output for the period 1995–1998

Category	Output	Category	Output
Energy & Fuels	405	Rehabilitation	46
Geosciences, Interdisciplinary	758	Sport Sciences	88
Engineering, Petroleum	14	Instruments & Instrumentation	589
Gastroenterology & Hepatology	2216	Geochemistry & Geophysics	409
Radiology, Nuclear Medicine & Medical Imaging	702	Mineralogy	215
Mathematics	1705	Engineering, Environmental	208
Medicine, General & Internal	4074	Computer Science, Hardware & Architecture	135
Education, Scientific Disciplines	127	Developmental Biology	296
Medical Informatics	63	Andrology	31
Entomology	268	Engineering, Biomedical	192
Chemistry, Multidisciplinary	2109	Oceanography	303
Construction & Building Technology	152	Psychology	645
Materials Science, Multidisciplinary	2311	Nuclear Science & Technology	595
Computer Science, Theory & Methods	565	Chemistry, Applied	1139
Computer Science, Information Systems	147	Physics, Fluids & Plasmas	512
Computer Science, Software, Graphics, Programming	198	Information Science & Library Science	100
Mathematics, Applied	1431	Physics, Particles & Fields	871
Acoustics	137	Chemistry, Medicinal	484
Cardiac & Cardiovascular Systems	1344	Materials Science, Paper & Wood	42

Table 7.33. (Cont.)

Category	Output	Category	Output
Respiratory System	525	Ergonomics	23
Agriculture, Dairy & Animal Science	426	Spectroscopy	646
Agriculture	1339	Operations Research & Management Science	260
Agriculture, Soil Science	419	Engineering, Marine	35
Nutrition & Dietetics	554	Thermodynamics	176
Food Science & Technology	1946	Materials Science, Coatings & Films	217
Anesthesiology	236	Fisheries	296
Anatomy & Morphology	287	Limnology	80
Engineering, Aerospace	48	Mathematics, Miscellaneous	93
Astronomy & Astrophysics	1897	Geography	103
Biochemistry & Molecular Biology	5131	Ornithology	176
Plant Sciences	2300	Materials Science, Biomaterials	98
Zoology	813	Geology	195
Biology	624	Computer Science, Interdisciplinary Applications	427
Biotechnology & Applied Microbiology	1671	Horticulture	171
Surgery	1936	Computer Science, Cybernetics	67
Crystallography	558	Transplantation	788
Biochemical Research Methods	1026	Psychology, Experimental	209
Cell Biology	1828	Psychology, Mathematical	29
Dermatology & Venereal Diseases	837	Materials Science, Characterization & Testing	40
Endocrinology & Metabolism	1386	Engineering, Civil	297
Hematology	1786	Mining & Mineral Processing	87
Environmental Sciences	1395	Physics, Mathematical	822
Water Resources	423	Electrochemistry	383
Marine & Freshwater Biology	1268	Psychology, Developmental	31
Metallurgy & Metallurgical Engineering	579	Education & Educational Research	71
Mechanics	426	Engineering, Industrial	45
Medicine, Research & Experimental	621	Mycology	284
Neurosciences	2793	Psychology, Educational	29
Clinical Neurology	1650	Environmental Studies	56
Pathology	915	Social Sciences, Biomedical	23
Obstetrics & Gynecology	480	Psychology, Biological	173
Dentistry, Oral Surgery & Medicine	424	Sociology	52
Ecology	865	Law	19
Oncology	1538	Microscopy	115
Ophthalmology	649	Remote Sensing	80

Table 7.33. (Cont.)

Category	Output	Category	Output
Orthopedics	242	Transportation	28
Otorhinolaryngology	129	Psychology, Clinical	133
Pediatrics	658	Imaging Science & Photographic Technology	66
Pharmacology & Pharmacy	2709	Materials Science, Textiles	47
Physics, Multidisciplinary	2220	Economics	480
Physiology	1208	Social Sciences, Mathematical Methods	113
Polymer Science	1027	Business	17
Physics, Applied	1502	Management	138
Engineering, Chemical	1163	Urban Studies	17
Psychiatry	495	Social Sciences, Interdisciplinary	37
Tropical Medicine	75	Planning & Development	47
Parasitology	302	Business, Finance	32
Veterinary Sciences	975	Social Issues	12
Virology	478	Public Administration	4
Substance Abuse	105	Social Work	12
Engineering, Mechanical	202	Psychology, Social	65
Microbiology	2004	Nursing	26
Statistics & Probability	446	Women's Studies	10
Physics, Nuclear	710	Area Studies	14
Optics	961	Political Science	42
Physics, Atomic, Molecular & Chemical	1110	Criminology & Penology	6
Biophysics	1066	International Relations	21
Chemistry, Organic	2653	Education, Special	9
Chemistry, Analytical	3073	Archeology	53
Medical Laboratory Technology	327	History	540
Chemistry, Physical	3464	Family Studies	7
Materials Science, Composites	26	History Of Social Sciences	24
Reproductive Systems	408	Psychology, Psychoanalysis	13
Genetics & Heredity	1740	Language & Linguistics	213
Immunology	2263	Psychology, Applied	38
Engineering, Electrical & Electronic	1466	Health Policy & Services	43
Chemistry, Inorganic & Nuclear	1732	Philosophy	121
Physics, Condensed Matter	2147	Ethnic Studies	3
Automation & Control Systems	176	Industrial Relations & Labor	8
Meteorology & Atmospheric Sciences	315	Communication	35
Behavioral Sciences	213	Demography	17
Toxicology	532	Engineering, Geological	22
Geriatrics & Gerontology	123	Health Care Sciences & Services	42
Engineering	185	Literature, Romance	321

Table 7.33. (Cont.)

Category	Output	Category	Output
Agricultural Economics & Policy	13	Theater	44
Forestry	204	Literary Reviews	97
History & Philosophy Of Science	97	Music	15
Computer Science, Artificial Intelligence	386	Literature	99
Engineering, Manufacturing	33	Poetry	8
Infectious Diseases	941	Literature, American	1
Rheumatology	627	Arts & Humanities, General	671
Urology & Nephrology	1746	Architecture	15
Telecommunications	147	Art	94
Paleontology	180	Religion	130
Allergy	606	Folklore	34
Biology, Miscellaneous	425	Film, Radio, Television	6
Materials Science, Ceramics	265	Classics	34
Public, Environmental & Occupational Health	587	Asian Studies	4
Emergency Medicine & Critical Care	212	Literature, German, Netherlandic, Scandinavian	1
Medicine, Legal	104	Literature, British Isles	17
Anthropology	99	Literature, African, Australian, Canadian	2
Peripheral Vascular Disease	845		

The number of categories that constitute this period is 223, none appearing in isolation.

Table 7.34. Listing of the JCR categories and Spain's output for the period 1999–2002

Category	Output	Category	Output
Energy & Fuels	589	Andrology	26
Geosciences, Interdisciplinary	1069	Engineering, Biomedical	318
Engineering, Petroleum	26	Oceanography	550
Gastroenterology & Hepatology	2504	Psychology	1026
Radiology, Nuclear Medicine & Medical Imaging	1059	Nuclear Science & Technology	620
Mathematics	2357	Chemistry, Applied	1776
Medicine, General & Internal	3718	Physics, Fluids & Plasmas	675
Education, Scientific Disciplines	115	Information Science & Library Science	100
Medical Informatics	108	Physics, Particles & Fields	1466
Entomology	364	Chemistry, Medicinal	673
Chemistry, Multidisciplinary	2391	Materials Science, Paper & Wood	66

Table 7.34. (Cont.)

Category	Output	Category	Output
Construction & Building Technology	205	Ergonomics	21
Materials Science, Multidisciplinary	3195	Spectroscopy	785
Computer Science, Theory & Methods	1030	Operations Research & Management Science	430
Computer Science, Information Systems	287	Engineering, Marine	30
Computer Science, Software, Graphics, Programming	408	Thermodynamics	283
Mathematics, Applied	2298	Materials Science, Coatings & Films	290
Acoustics	191	Fisheries	503
Cardiac & Cardiovascular Systems	2318	Limnology	137
Respiratory System	980	Mathematics, Miscellaneous	147
Agriculture, Dairy & Animal Science	487	Geography	117
Agriculture	1168	Ornithology	216
Agriculture, Soil Science	447	Materials Science, Biomaterials	176
Nutrition & Dietetics	858	Geology	247
Food Science & Technology	2688	Computer Science, Interdisciplinary Applications	757
Anesthesiology	249	Horticulture	297
Anatomy & Morphology	298	Computer Science, Cybernetics	93
Engineering, Aerospace	87	Transplantation	1057
Astronomy & Astrophysics	2312	Psychology, Experimental	359
Biochemistry & Molecular Biology	6356	Psychology, Mathematical	38
Plant Sciences	2482	Materials Science, Characterization & Testing	58
Zoology	945	Engineering, Civil	248
Biology	726	Mining & Mineral Processing	108
Biotechnology & Applied Microbiology	2081	Physics, Mathematical	1176
Surgery	2167	Electrochemistry	480
Crystallography	708	Psychology, Developmental	52
Biochemical Research Methods	1446	Education & Educational Research	98
Cell Biology	2265	Engineering, Industrial	110
Dermatology & Venereal Diseases	769	Mycology	376
Endocrinology & Metabolism	1760	Psychology, Educational	35

Table 7.34. (Cont.)

Category	Output	Category	Output
Hematology	2045	Environmental Studies	160
Environmental Sciences	1931	Social Sciences, Biomedical	52
Water Resources	655	Psychology, Biological	206
Marine & Freshwater Biology	1526	Sociology	62
Metallurgy & Metallurgical Engineering	732	Law	22
Mechanics	562	Microscopy	117
Medicine, Research & Experimental	818	Remote Sensing	102
Neurosciences	3492	Transportation	30
Clinical Neurology	3045	Psychology, Clinical	137
Pathology	1181	Imaging Science & Photographic Technology	81
Obstetrics & Gynecology	656	Materials Science, Textiles	90
Dentistry, Oral Surgery & Medicine	362	Economics	824
Ecology	1255	Social Sciences, Mathematical Methods	192
Oncology	2116	Business	103
Ophthalmology	578	Management	186
Orthopedics	241	Urban Studies	51
Otorhinolaryngology	186	Social Sciences, Interdisciplinary	54
Pediatrics	609	Planning & Development	62
Pharmacology & Pharmacy	3144	Business, Finance	57
Physics, Multidisciplinary	2187	Social Issues	20
Physiology	881	Public Administration	18
Polymer Science	1301	Social Work	12
Physics, Applied	1898	Psychology, Social	91
Engineering, Chemical	1739	Nursing	31
Psychiatry	1092	Women's Studies	13
Tropical Medicine	90	Area Studies	12
Parasitology	398	Political Science	48
Veterinary Sciences	993	Criminology & Penology	3
Virology	733	International Relations	34
Substance Abuse	139	Education, Special	17
Engineering, Mechanical	377	Archeology	71
Microbiology	2892	History	662
Statistics & Probability	695	Family Studies	16
Physics, Nuclear	817	History Of Social Sciences	42
Optics	1503	Psychology, Psychoanalysis	33

Table 7.34. (Cont.)

Category	Output	Category	Output
Physics, Atomic, Molecular & Chemical	1646	Language & Linguistics	287
Biophysics	1140	Psychology, Applied	82
Chemistry, Organic	3015	Health Policy & Services	101
Chemistry, Analytical	3524	Philosophy	155
Medical Laboratory Technology	406	Ethnic Studies	4
Chemistry, Physical	4398	Industrial Relations & Labor	13
Materials Science, Composites	134	Communication	33
Reproductive Systems	610	Demography	19
Genetics & Heredity	2024	Engineering, Geological	59
Immunology	2945	Health Care Sciences & Services	177
Engineering, Electrical & Electronic	2276	Literature, Romance	390
Chemistry, Inorganic & Nuclear	2208	Theater	25
Physics, Condensed Matter	2756	Literary Reviews	68
Automation & Control Systems	306	Engineering, Ocean	18
Meteorology & Atmospheric Sciences	545	Music	19
Behavioral Sciences	399	Literature	91
Toxicology	659	Poetry	1
Geriatrics & Gerontology	223	Literature, American	3
Engineering	335	Arts & Humanities, General	587
Agricultural Economics & Policy	26	Architecture	21
Forestry	295	Art	110
History & Philosophy Of Science	131	Religion	139
Computer Science, Artificial Intelligence	1082	Folklore	55
Engineering, Manufacturing	102	Film, Radio, Television	3
Infectious Diseases	1631	Classics	38
Rheumatology	537	Asian Studies	15
Urology & Nephrology	1468	Literature, German, Netherlandic, Scandinavian	2
Telecommunications	250	Literature, British Isles	14
Paleontology	257	Literature, African, Australian, Canadian	6
Allergy	736	Agriculture, Multidisciplinary	426
Biology, Miscellaneous	572	Robotics	54
Materials Science, Ceramics	752	Ethics	22
Public, Environmental & Occupational Health	898	Transportation Science & Technology	42

Table 7.34. (Cont.)

Category	Output	Category	Output
Emergency Medicine & Critical Care	262	Applied Linguistics	73
Medicine, Legal	169	Integrative & Complementary Medicine	18
Anthropology	170	Biodiversity Conservation	118
Peripheral Vascular Disease	1339	Psychology, Multidisciplinary	217
Rehabilitation	73	Literary Theory & Criticism	150
Sport Sciences	171	Critical Care Medicine	330
Instruments & Instrumentation	684	Gerontology	27
Geochemistry & Geophysics	644	Medical Ethics	9
Mineralogy	248	Neuroimaging	50
Engineering, Environmental	550	Agricultural Engineering	95
Computer Science, Hardware & Architecture	218	Evolutionary Biology	128
Developmental Biology	426	Geography, Physical	60

The scientogram of this latest period (1992–2002) contains 240 categories, none of them isolated from the rest.

The three scientograms present the familiar structure seen throughout this book: a neuron with a large central neurite from which several prolongations extend outward. A comparative look at the three reveals that the more recent the period of study, the larger the central neurite and the more uniform and reduced the number of prolongations it has. Hence, evolution is one of consistent advancement and increase in the categories that make up the focus of Spanish research overall. At the same time, we corroborate a better definition and unification of the main lines of investigation within this geographic domain.

For a more detailed study of this domain's evolution, we perform FA on the original cocitation matrix of each period. Then we transfer the results to each of the corresponding scientograms.

7.3.2 Factor Analysis

In agreement with the procedure explained in Sect. 7.1.2., we extracted and categorized 10 factors from the period 1990–1994 (Table 7.35), another 10 from 1995–1998 (Table 7.36), and 11 from 1999–2002 (Table 7.37). The name of each factor, along with its individual and accumulated variance, is given in the tables.

The tables with the categories comprising each of the factors extracted for each time period, together with the corresponding factor loadings, are reunited up in Annex IV.

Table 7.35. Factors extracted from Spain's scientific output domain for 1990–1994

Factor	Factor name	% variance	% of accumulative variance
1	Biomedicine	23.1	23.1
2	Materials Sciences and Applied Physics	12.9	36
3	Psychology	7.8	43.8
4	Agriculture and Soil Sciences	6.3	50.1
5	Computer Science and Telecommunications	4.1	54.2
6	Earth & Spaces Sciences	3.1	57.3
7	Management, Law and Economics	3.3	60.6
8	Arts & Humanities	2.4	63
9	Animal Biology, Ecology	2.4	65.4
10	Nuclear Physics and Particle Physics	1.5	66.9

Table 7.36. Factors extracted from the Spanish geographic domain for 1995–1998

Factor	Factor's name	% variance	% of accumulative variance
1	Biomedicine	23.1	23.1
2	Materials Sciences and Applied Physics	13.2	36.3
3	Psychology	8.2	44.5
4	Agriculture and Soil Sciences	6.8	51.3
5	Computer Science and Telecommunications	4.4	55.7
6	Earth & Spaces Sciences	4	59.7
7	Management, Law and Economics	3.4	63.1
8	Humanities	3.1	66.2
9	Animal Biology, Ecology	2.3	68.5
10	Nuclear Physics and Particle Physics	1.4	69.9

An analysis of Tables 7.35–7.37 shows the two first time periods to be practically identical, with just a slight variation in the denomination and ordering of a couple of factors, and in the cumulative variance of one or two. Differences with the third period are greater. We can see a descending movement, or a loss of interest on the part of researchers of the domain, in those factors related with *Agriculture* and *Social Sciences,* and *Humanities.* In contrast, there is an upward movement or growth of interest in areas

Table 7.37. Factors extracted from the domain Spanish scientific output, 1999–2002

Factor	Factor's name	% variance	% of accumulative variance
1	Biomedicine	22.1	22.1
2	Materials Sciences and Applied Physics	12.3	34.4
3	Psychology	7.7	42.1
4	Agriculture and Soil Sciences	6.9	49
5	Computer Science and Telecommunications	4.4	53.4
6	Earth & Spaces Sciences	4.1	57.5
7	Management, Law and Economics	3.4	60.9
8	Animal Biology, Ecology	2.7	63.6
9	Política Sanitaria y Servicios Médicos	2.1	65.7
10	Humanities	1.9	67.6
11	Nuclear Physics and Particle Physics	1.7	69.3

related with *Technology* and *Environment* and *Mental Health*, where some new factors even appear on the scene.

Factor Scientograms

The factor scientogram for the period 1990–1994 leaves 26 categories unfactorized (Fig. 7.23); the period 1995–1998 leaves behind 36 (Fig. 7.24); and the final period 1999–2002 lets 29 go without factorizing (Fig. 7.25), as shown in Table 7.38.

Nonetheless, out of the top 26 categories for the period 1990–1994, only seven continue to go unfactorized by 1999–2002 (Table 7.39).

This signals the fact that those 26 categories participating least in Spanish research during the period 1990–1994 (confirmed by their low factor loading shown in Annex IV) eventually evolve to the point where only seven categories continue to be unfactorized 12 years later. The rest, as a consequence of research efforts, gain weight in the Spanish realm over the years. Yet the development of scientific research makes new categories arise as well, though for the time being they have very limited relevance in the scientific panorama of Spain. This explains the high number of uncategorized factors seen for the period 1999–2002.

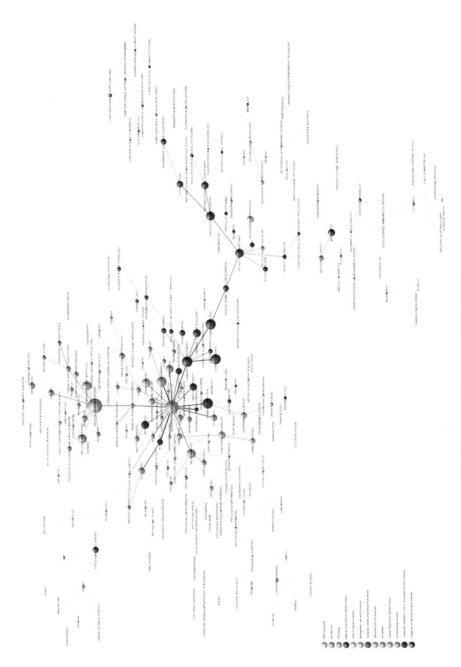

Fig. 7.23. Factor scientogram of Spain's scientific output, 1990–1994

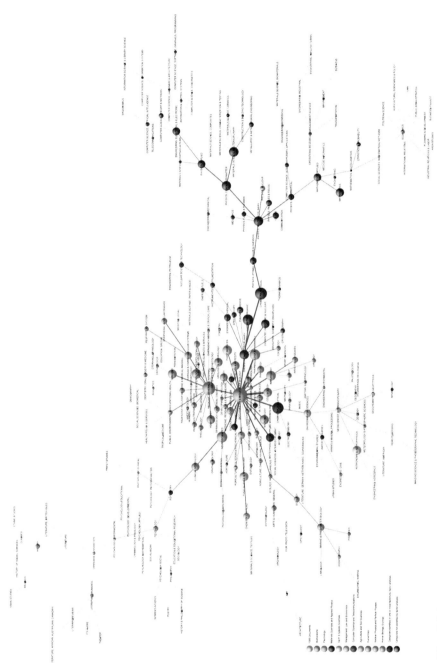

Fig. 7.24. Factor scientogram of the domain of Spain's scientific output, 1995–1998

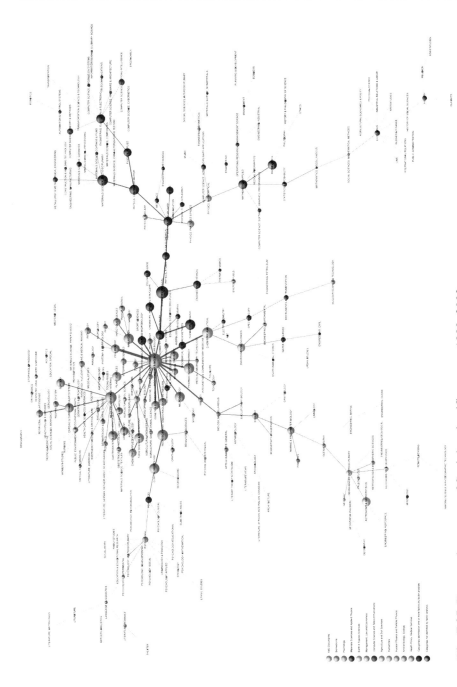

Fig. 7.25. Factor scientogram of Spain's scientific output, 1999–2002

Table 7.38. Unfactorized categories in each one of the three periods of study for the Spanish domain

1990–1994	1995–1998	1999–2002
Acoustics	Archeology	Acoustics
Anthropology	Architecture	Architecture
Archeology	Art	Art
Architecture	Business, Finance	Computer Science, Interdisciplinary Applications
Area Studies	Classics	Economics
Arts & Humanities, General	Computer Science, Theory & Methods	Engineering
Asian Studies	Construction & Building Technology	Engineering, Civil
Construction & Building Technology	Demography	Engineering, Industrial
Demography	Engineering	Environmental Studies
Engineering	Engineering, Manufacturing	Ethnic Studies
Engineering, Civil	Engineering, Mechanical	Folklore
Engineering, Mechanical	Ethnic Studies	History
Environmental Studies	Geology	History & Philosophy Of Science
Film, Radio, Televisión	Health Care Sciences & Services	History Of Social Sciences
Health Policy & Services	Health Policy & Services	Literary Theory & Criticism
History & Philosophy Of Science	History & Philosophy Of Science	Literature, African, Australian, Canadian
Literature, German, Netherlandic, Scandinavian	Horticulture	Literature, British Isles
Materials Science, Biomaterials	Literature, British Isles	Literature, German, Netherlandic, Scandinavian
Materials Science, Textiles	Literature, German, Netherlandic, Scandinavian	Management
Mathematics	Materials Science, Biomaterials	Mathematics
Mathematics, Miscellaneous	Mathematics	Mathematics, Applied
Nursing	Mathematics, Applied	Mathematics, Miscellaneous
Philosophy	Mathematics, Miscellaneous	Medical Informatics

Table 7.38. (Cont.)

1990–1994	1995–1998	1999–2002
Poetry	Medical Informatics	Music
Social Sigues	Nursing	Statistics & Probability
Statistics & Probability	Ornithology	Theater
	Philosophy	Transportation
	Physics, Particles &	Transportation Science &
	Fields	Technology
	Poetry	Urban Studies
	Polymer Science	
	Psychology, Social	
	Religión	
	Social Sigues	
	Social Sciences,	
	Biomedical	
	Social Sciences,	
	Interdisciplinary	
	Statistics & Probability	

Table 7.39. Categories consistently unfactorized over the three time periods of study for the Spanish domain

1990–2002
Architecture
Engineering
History & Philosophy of Science
Literature, German, Netherlandic, Scandinavian
Mathematics
Mathematics, Miscellaneous
Statistics & Probability

7.3.3 Evolution of a Domain

Macrostructure

The scientograms of the three periods (Figs. 7.26–7.28) show the classical bibliometric distribution of a few large thematic areas surrounded by a great many small ones. At the same time, there is a major central thematic area that acts as the point of interconnection of other smaller thematic areas around it. As we advance in time over the factor scientograms, we reconfirm that the number of categories occupying central positions is greater – as we saw earlier on, when looking at the non-factor scientograms. This does not entail an increase in the number of categories in the area of *Biomedicine*, which is practically unchanging over the three periods (see Annex IV). Rather, it means that over time, categories from other areas, such as *Animal Biology and Ecology* or *Agriculture and Soil Sciences*, have come to adopt

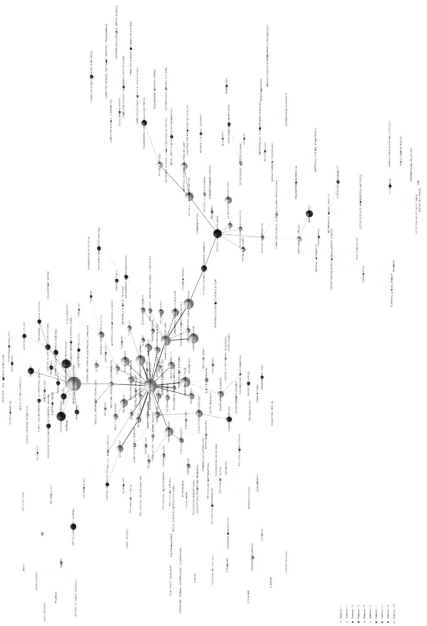

Fig. 7.26. Scientogram of distances of the Spanish domain, 1990–1994, with respect to the central category

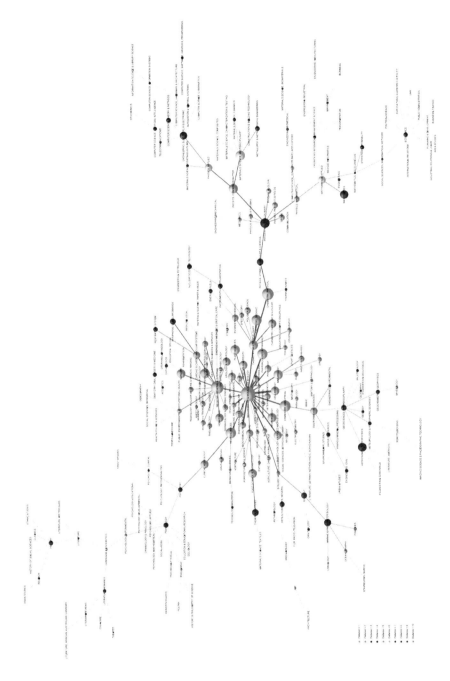

Fig. 7.27. Scientogram of distances of Spain's scientific output, 1995–1998, with respect to the central category

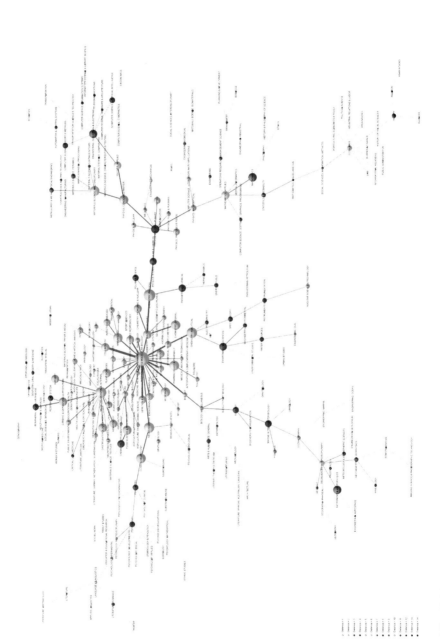

Fig. 7.28. Scientogram of distances of the Spanish scientific domain 1999–2002, with respect to the central category

important positions in the scientograms by sharing sources with categories that were around it in previous years. In this way they manage to heighten their universality and the interest of other researchers in that research.

Unlike other domains, Spain makes it difficult to reproduce what we have been typifying as the scheme of macrostructural vertebration of science in developed nations. That is, the situation of the Biomedical and Earth Sciences in the central zone, the hard sciences to the right, and the softer sciences to the left. In general, the Spanish domain follows this pattern to the center and right, but not toward the left. In our domain, the categories making up *Mathematics* do not bridge the social sciences and humanities over to the rest of the sciences, leaving them disperse in the different scientograms. Hence, the area *Management, Law and Economics* appears to the right, joined to *Mathematics*, rather than to the left.

The relative positions taken up by the thematic areas are essentially the same over all this time of study. The only difference is apparent in the period 1990–1994, where *Earth and Space Sciences* is situated in the right zone, whereas they appear in the central zone of the other two depictions.

Centrality

The networks of thematic areas constructed from each of the factor scientograms and the paths that connect their contents allow us to easily calculate the degree of centrality of each thematic area for each of the three periods (Table 7.40). Thus:

Table 7.40. Degree centrality of the thematic areas of the Spanish scientific domain

1990–1994		1995–1998	
Thematic area	Grade	Thematic area	Grade
Biomedicine	4	Biomedicine	4
Materials Sciences and Applied Physics	4	Materials Sciences and Applied Physics	4
Animal Biology, Ecology	2	Animal Biology, Ecology	2
Nuclear Physics and Particle Physics	2	Nuclear Physics and Particle Physics	2
Agriculture and Soil Sciences	1	Agriculture and Soil Sciences	1
Arts and Humanities	1	Earth & Space Sciences	1
Earth & Space Sciences	1	Management, Law and Economics	1
Management, Law and Economics	1	Humanities	1
Computer Science and Telecommunications	1	Computer Science and Telecommunications	1
Psychology	1	Psychology	1

Table 7.40. (Cont.)

1999–2002	
Thematic Area	Grade
Biomedicine	5
Materials Sciences and Applied Physics	3
Animal Biology, Ecology	2
Nuclear Physics and Particle Physics	2
Psychology	2
Agriculture and Soil Sciences	1
Earth & Spaces Sciences	1
Management, Law and Economics	1
Humanities	1
Computer Science and Telecommunications	1
Health Policy, Medical Services	1

In the periods 1990–1994 and 1995–1998, the two areas with the most degree centrality are *Biomedicine* and *Materials Sciences & Applied Physics*. On the other hand, for the period 1999–2002 there is only one, outstanding in its centrality: the area of *Biomedicine*. In other words, in the first two periods, *Biomedicine* for one, and in second place *Materials Science and Applied Physics* are, on an equal level, the most universal thematic areas of the domain. Yet the final 4-year period witnesses a new exclusive protagonist under the name *Biomedicine*.

Prominence

The case of prominence is much the same. The grade robbery algorithm, in charge of detecting the most prominent thematic areas in each temporal space, identifies the same thematic areas as does centrality. See the tables below (Table 7.41) for a synopsis.[6]

Table 7.41. Prominence of the thematic areas of the Spanish scientific domain

1990–1994		1995–1998	
Thematic area	Grade	Thematic area	Grade
Biomedicine	9	Biomedicine	9
Materials Sciences and Applied Physics	9	Materials Sciences and Applied Physics	9

1999–2002	
Thematic area	Grade
Biomedicine	20

---

[6] The thematic areas not appearing had a zero value.

Paths of Interconnection Among Thematic Areas

At the risk of repeating ourselves, the relative positions of the thematic areas are modified very little over the 12 years comprised in our analysis. Yet the ways in which they are interconnected with each other varies substantially. In the first two periods there are important differences in terms of the paths of interconnection among thematic areas.

In the period 1990–1994, *Psychology* seems to have an orientation more geared toward the study of the mind, from whence its path of connection with *Biomedicine*: *Biochemistry & Molecular Biology* ←→ *Neurosciences* ←→ *Psychology*. In the intervals 1995–1998 and 1999–2002, research in this area is reoriented to revolve around clinical and pathological studies, as the veering of its route reveals: *Biochemistry & Molecular Biology* ←→ *Neurosciences* ←→ *Clinical Neurology* ←→ *Psychiatry* ←→ *Psychology*.

In the first of these periods, *Humanities* appears connected with *Animal Biology and Ecology* through *Anthopology* ←→ *Biology Miscellaneous* ←→ *Biology*; while in the other two they are linked to *Psychology* over the category *Language and Linguistics*, in the following manner: *History* ←→ *Literature* ←→ *Language & Linguistics* ←→ *Psychology Experimental* ←→ *Psychology*. This demonstrates a change in research goals and in the sources used in this particular area.

The thematic area of *Earth and Space Sciences* also undergoes modifications over this period of study. Between 1990 and 1994, it appears linked to *Materials Sciences and Applied Physics*: *Geosciences Interdisciplinary* ←→ *Astronomy & Astrophysics* ←→ *Physics Multidisciplinary*. In the space of 1995–1998, its strongest connection is established with *Biomedicine*: *Geosciences Interdisciplinary* ←→ *Environmental Sciences* ←→ *Chemistry Analytical* ←→ *Biochemistry & Molecular Biology*. Finally, in the period 1999–2002, we see its basic research divided into the thematic areas of *Animal Biology & Ecology*: *Geosciences*, ←→ *Oceanography* ←→ *Marine & Freshwater Biology* ←→ *Ecology* ←→ *Biology Miscellaneous* ←→ *Biology*.

Points of Interaction Between Adjacent Thematic Areas

As we have indicated more than once, the red categories in the scientograms signal interchange, and if we now look at the red hot categories of the first period, we will begin to perceive the evolution of Spanish research over time.

Table 7.42 shows that 15 categories produced interactive movement of sources between different thematic areas as a consequence of the curiosity of researchers during that time. The fundamental pairings of thematic areas

Table 7.42. Multithematic categories of the Spanish scientific domain, 1990–1994

Category	Thematic areas	
Biochemical Research Methods	Biomedicine	Agriculture and Soil Sciences
Biophysics	Biomedicine	Agriculture and Soil Sciences
Biotechnology & Applied Microbiology	Biomedicine	Agriculture and Soil Sciences
Chemistry, Analytical	Biomedicine	Agriculture and Soil Sciences
Engineering, Manufacturing	Materials Sciences and Applied Physics	Computer Science and Telecommunications
Ergonomics	Psychology	Computer Science and Telecommunications
Genetics & Heredity	Biomedicine	Agriculture and Soil Sciences
Management	Computer Science and Telecommunications	Management, Law and Economics
Microbiology	Biomedicine	Agriculture and Soil Sciences
Mineralogy	Materials Sciences and Applied Physics	Nuclear Physics and Particle Physics
Oceanography	Earth & Spaces Sciences	Animal Biology, Ecology
Optics	Materials Sciences and Applied Physics	Nuclear Physics and Particle Physics
Physics, Fluids & Plasmas	Materials Sciences and Applied Physics	Nuclear Physics and Particle Physics
Physics, Mathematical	Materials Sciences and Applied Physics	Nuclear Physics and Particle Physics
Physics, Nuclear	Materials Sciences and Applied Physics	Nuclear Physics and Particle Physics

are two: *Biomedicine* with *Agriculture and Earth Sciences* and *Materials Sciences & Applied Physics* with *Nuclear and Particle Physics*.

In the above period 1995–1998, there are 19 areas involved in intellectual interchange (Table 7.43). Eight of these are in bold, shown in Table 7.43 with their thematic partners, coinciding with couplings during the period 1990–1994. Some of the 11 new categories that arise the continuation of interchanges from the previous period, although others inaugurate new lines of research by means of the pairing of areas *Biomedicine* and *Psychology*.

Table 7.43. Multithematic categories of the Spanish scientific domain, 1995–1998

Category	Thematic areas	
Biochemical Research Methods	*Biomedicine*	*Agriculture & Soil Sciences*
Biotechnology & Applied Microbiology	*Biomedicine*	*Agriculture & Soil Sciences*
Chemistry, Analytical	*Biomedicine*	*Agriculture & Soil Sciences*
Management	*Computer Science and Telecommunications*	*Management, Law and Economics*
Mineralogy	*Materials Sciences and Applied Physics*	*Earth & Spaces Sciences*
Optics	*Materials Sciences and Applied Physics*	*Nuclear Physics and Particle Physics*
Physics, Fluids & Plasmas	*Materials Sciences and Applied Physics*	*Nuclear Physics and Particle Physics*
Physics, Mathematical	*Materials Sciences and Applied Physics*	*Nuclear Physics and Particle Physics*
Agricultural Economics & Policy	Management, Law and Economics	Agriculture and Soil Sciences
Communication	Materials Sciences and Applied Physics	Nuclear Physics and Particle Physics
Geography	Earth & Spaces Sciences	Animal Biology, Ecology
Mycology	Biomedicine	Agriculture and Soil Sciences
Nuclear Science & Technology	Materials Sciences and Applied Physics	Nuclear Physics and Particle Physics
Nutrition & Dietetics	Biomedicine	Agriculture and Soil Sciences
Psychiatry	Biomedicine	Psychology
Rehabilitation	Biomedicine	Psychology
Substance Abuse	Biomedicine	Psychology
Water Resources	Earth & Spaces Sciences	Agriculture and Soil Sciences
Zoology	Biomedicine	Animal Biology, Ecology

To finish with this evolutive incursion, in the interval 1999–2002, we find 16 categories responsible for the flow of information between different thematic areas (Table 7.44). Five of them, in bold, are common to the two anterior periods and put the very same thematic areas in contact. Six, shown in italics, are repeated from the previous period alone, and also relate the same thematic periods as before. And finally, the last five show what might be considered as a new coupling and therefore the beginnings of a new nest of research: *Agriculture and Soil Sciences* with *Biology Animal and Ecology*.

Table 7.44. Multithematic categories of the Spanish scientific domain, 1999–2002

Category	Thematic areas	
Biochemical Research Methods	**Biomedicine**	**Agriculture and Soil Sciences**
Biotechnology & Applied Microbiology	**Biomedicine**	**Agriculture and Soil Sciences**
Mineralogy	**Materials Sciences and Applied Physics**	**Earth & Spaces Sciences**
Optics	**Materials Sciences and Applied Physics**	**Nuclear Physics and Particle Physics**
Physics, Fluids & Plasmas	**Materials Sciences and Applied Physics**	**Nuclear Physics and Particle Physics**
Water Resources	*Earth & Spaces Sciences*	*Agriculture and Soil Sciences*
Substance Abuse	*Biomedicine*	*Psychology*
Psychiatry	*Biomedicine*	*Psychology*
Nutrition & Dietetics	*Biomedicine*	*Agriculture and Soil Sciences*
Geography	*Earth & Spaces Sciences*	*Biología Animal y Ecología*
Communication	*Materials Sciences and Applied Physics*	*Nuclear Physics and Particle Physics*
Entomology	Agriculture and Soil Sciences	Animal Biology, Ecology
Forestry	Agriculture and Soil Sciences	Animal Biology, Ecology
Health Care Sciences & Services	Biomedicine	Health Policy, Medical Services
Mining & Mineral Processing	Materials Sciences and Applied Physics	Earth & Spaces Sciences
Women's Studies	Psychology	Health Policy, Medical Services

Microstructure

Returning again to our Darwinian scientograms, and more specifically to the factorial displays, we see that on a microstructural level there is a reproduction of macrostructural characteristics. In terms of categories, the three scientograms offer a clear example of the hyperbolic bibliometric tradition. Similarly, in all three cases a greater concentration of categories takes place in the central and right areas, while respecting the central–peripheral structure as well.

As inferred from Table 7.45, *Biochemistry & Molecular Biology* is the most central category of all three temporal scientograms. The key role of

Table 7.45. Degree centrality of the top 15 categories of the Spanish domain in its three periods of study

1990–1994		1995–1998	
Category	Grade	Category	Grade
Biochemistry & Molecular Biology	26	Biochemistry & Molecular Biology	26
Medicine, General & Internal	17	Medicine, General & Internal	22
Psychology	14	Psychology	12
Chemistry, Multidisciplinary	9	Economics	10
Physics, Multidisciplinary	9	Geosciences, Interdisciplinary	9
Geosciences, Interdisciplinary	7	Chemistry, Multidisciplinary	9
Engineering, Electrical & Electronic	7	Physics, Multidisciplinary	9
Economics	7	Engineering, Electrical & Electronic	7
Mathematics, Applied	6	Materials Science, Multidisciplinary	6
Agriculture	6	Mathematics, Applied	6
Neurosciences	6	Agriculture	6
Materials Science, Multidisciplinary	5	Environmental Sciences	5
Chemistry, Analytical	5	Public, Environmental & Occupational Health	5
History	5	History	5
Language & Linguistics	5	Surgery	4

1999–2002	
Category	Grade
Biochemistry & Molecular Biology	26
Medicine, General & Internal	20
Psychology	13
Economics	10
Materials Science, Multidisciplinary	9
Physics, Multidisciplinary	9
Engineering, Electrical & Electronic	9
Geosciences, Interdisciplinary	8
Agriculture	8
Mathematics, Applied	7
Chemistry, Multidisciplinary	6
Psychiatry	5
Chemistry, Analytical	5
Immunology	5
Biology, Miscellaneous	5

this category is now clear: it has been shown, visualization after visualization, to be the most universal, and the fact that its nodal grade is the same in all three temporal analyses avows for the permanence of its protagonistic status.

The degree of universality and participation that categories wield in the scientific development of each period is associated with the geodesic distance of each with respect to the star category of *Biochemistry & Molecular Biology* (Tables 7.46–7.48).

Table 7.46. Distances of the categories of the scientogram of Spain's scientific output, 1990–1994 with respect to the central category

Category	Distance	Category	Distance
Biochemistry & Molecular Biology	0	Social Issues	3
Chemistry, Multidisciplinary	1	Social Work	3
Agriculture	1	Psychology, Social	3
Nutrition & Dietetics	1	Nursing	3
Plant Sciences	1	Criminology & Penology	3
Zoology	1	Psychology, Psychoanalysis	3
Biology	1	Psychology, Applied	3
Biotechnology & Applied Microbiology	1	Literature, African, Australian, Canadian	3
Biochemical Research Methods	1	Engineering, Petroleum	4
Endocrinology & Metabolism	1	Acoustics	4
Medicine, Research & Experimental	1	Respiratory System	4
Neurosciences	1	Dentistry, Oral Surgery & Medicine	4
Oncology	1	Orthopedics	4
Pharmacology & Pharmacy	1	Otorhinolaryngology	4
Physiology	1	Physics, Multidisciplinary	4
Parasitology	1	Tropical Medicine	4
Veterinary Sciences	1	Peripheral Vascular Disease	4
Virology	1	Nuclear Science & Technology	4
Microbiology	1	Engineering, Marine	4
Biophysics	1	Engineering, Civil	4
Chemistry, Analytical	1	Social Sciences, Biomedical	4
Reproductive Systems	1	Urban Studies	4
Genetics & Heredity	1	Women's Studies	4
Immunology	1	Political Science	4
Geriatrics & Gerontology	1	Education, Special	4
Developmental Biology	1	Archeology	4
Chemistry, Medicinal	1	Family Studies	4
Medicine, General & Internal	2	Language & Linguistics	4

Table 7.46. (Cont.)

Category	Distance	Category	Distance
Education, Scientific Disciplines	2	Health Policy & Services	4
Entomology	2	Demography	4
Agriculture, Dairy & Animal Science	2	Arts & Humanities, General	4
Agriculture, Soil Science	2	Astronomy & Astrophysics	5
Food Science & Technology	2	Mechanics	5
Crystallography	2	Physics, Nuclear	5
Cell Biology	2	Optics	5
Environmental Sciences	2	Physics, Condensed Matter	5
Marine & Freshwater Biology	2	Physics, Particles & Fields	5
Clinical Neurology	2	Physics, Mathematical	5
Ecology	2	History	5
Ophthalmology	2	Communication	5
Polymer Science	2	Literature, Romance	5
Engineering, Chemical	2	Literature	5
Psychiatry	2	Architecture	5
Chemistry, Organic	2	Art	5
Medical Laboratory Technology	2	Film, Radio, Television	5
Chemistry, Physical	2	Classics	5
Chemistry, Inorganic & Nuclear	2	Literature, British Isles	5
Behavioral Sciences	2	Geosciences, Interdisciplinary	6
Toxicology	2	Materials Science, Multidisciplinary	6
Forestry	2	Mathematics, Applied	6
Infectious Diseases	2	Physics, Applied	6
Allergy	2	Engineering, Mechanical	6
Biology, Miscellaneous	2	Physics, Fluids & Plasmas	6
Medicine, Legal	2	Computer Science, Interdisciplinary Applications	6
Sport Sciences	2	Area Studies	6
Andrology	2	History Of Social Sciences	6
Psychology	2	Theater	6
Spectroscopy	2	Literary Reviews	6
Ornithology	2	Music	6
Horticulture	2	Religion	6
Transplantation	2	Folklore	6
Electrochemistry	2	Mathematics	7
Mycology	2	Medical Informatics	7
Philosophy	2	Construction & Building Technology	7
Energy & Fuels	3	Metallurgy & Metallurgical Engineering	7

Table 7.46. (Cont.)

Category	Distance	Category	Distance
Gastroenterology & Hepatology	3	Engineering, Electrical & Electronic	7
Radiology, Nuclear Medicine & Medical Imaging	3	Meteorology & Atmospheric Sciences	7
Cardiac & Cardiovascular Systems	3	Engineering	7
Anesthesiology	3	Paleontology	7
Anatomy & Morphology	3	Materials Science, Ceramics	7
Surgery	3	Geochemistry & Geophysics	7
Dermatology & Venereal Diseases	3	Engineering, Biomedical	7
Hematology	3	Operations Research & Management Science	7
Water Resources	3	Mathematics, Miscellaneous	7
Pathology	3	Geography	7
Obstetrics & Gynecology	3	Geology	7
Pediatrics	3	Materials Science, Characterization & Testing	7
Substance Abuse	3	Mining & Mineral Processing	7
Physics, Atomic, Molecular & Chemical	3	Asian Studies	7
Materials Science, Composites	3	Engineering, Aerospace	8
History & Philosophy Of Science	3	Statistics & Probability	8
Rheumatology	3	Automation & Control Systems	8
Urology & Nephrology	3	Computer Science, Artificial Intelligence	8
Public, Environmental & Occupational Health	3	Engineering, Manufacturing	8
Emergency Medicine & Critical Care	3	Telecommunications	8
Anthropology	3	Mineralogy	8
Rehabilitation	3	Computer Science, Hardware & Architecture	8
Instruments & Instrumentation	3	Materials Science, Biomaterials	8
Engineering, Environmental	3	Computer Science, Cybernetics	8
Oceanography	3	Remote Sensing	8
Chemistry, Applied	3	Transportation	8
Materials Science, Paper & Wood	3	Social Sciences, Mathematical Methods	8
Ergonomics	3	Management	8
Thermodynamics	3	Computer Science, Theory & Methods	9

Table 7.46. (Cont.)

Category	Distance	Category	Distance
Materials Science, Coatings & Films	3	Computer Science, Information Systems	9
Fisheries	3	Computer Science, Software, Graphics, Programming	9
Limnology	3	Imaging Science & Photographic Technology	9
Psychology, Experimental	3	Economics	9
Psychology, Mathematical	3	Business	9
Psychology, Developmental	3	Planning & Development	9
Education & Educational Research	3	Agricultural Economics & Policy	10
Engineering, Industrial	3	Information Science & Library Science	10
Psychology, Educational	3	Law	10
Environmental Studies	3	Business, Finance	10
Psychology, Biological	3	Public Administration	10
Sociology	3	International Relations	10
Microscopy	3	Industrial Relations & Labor	10
Psychology, Clinical	3	Poetry	
Materials Science, Textiles	3	Literature, German, Netherlandic, Scandinavian	
Social Sciences, Interdisciplinary	3		

Table 7.47. Distances of the categories of the scientogram of Spain's scientific output, 1995–1998, with respect to the central category

Category	Distance	Category	Distance
Biochemistry & Molecular Biology	0	Meteorology & Atmospheric Sciences	4
Medicine, General & Internal	1	Paleontology	4
Chemistry, Multidisciplinary	1	Geochemistry & Geophysics	4
Agriculture	1	Psychology	4
Nutrition & Dietetics	1	Nuclear Science & Technology	4
Plant Sciences	1	Geography	4
Zoology	1	Ornithology	4
Biology	1	Geology	4
Biotechnology & Applied Microbiology	1	Engineering, Civil	4
Biochemical Research Methods	1	Mining & Mineral Processing	4
Cell Biology	1	Psychology, Clinical	4

Table 7.47. (Cont.)

Category	Distance	Category	Distance
Endocrinology & Metabolism	1	Materials Science, Textiles	4
Medicine, Research & Experimental	1	Urban Studies	4
Neurosciences	1	Archeology	4
Oncology	1	Psychology, Psychoanalysis	4
Pharmacology & Pharmacy	1	Engineering, Geological	4
Physiology	1	Film, Radio, Television	4
Veterinary Sciences	1	Engineering, Aerospace	5
Microbiology	1	Mechanics	5
Biophysics	1	Physics, Nuclear	5
Chemistry, Analytical	1	Optics	5
Medical Laboratory Technology	1	Physics, Condensed Matter	5
Reproductive Systems	1	Mineralogy	5
Genetics & Heredity	1	Oceanography	5
Immunology	1	Physics, Fluids & Plasmas	5
Developmental Biology	1	Physics, Particles & Fields	5
Chemistry, Medicinal	1	Fisheries	5
Gastroenterology & Hepatology	2	Limnology	5
Radiology, Nuclear Medicine & Medical Imaging	2	Psychology, Experimental	5
Education, Scientific Disciplines	2	Psychology, Mathematical	5
Entomology	2	Physics, Mathematical	5
Cardiac & Cardiovascular Systems	2	Psychology, Developmental	5
Agriculture, Dairy & Animal Science	2	Education & Educational Research	5
Agriculture, Soil Science	2	Psychology, Educational	5
Food Science & Technology	2	Sociology	5
Anesthesiology	2	Remote Sensing	5
Anatomy & Morphology	2	Social Work	5
Surgery	2	Psychology, Social	5
Crystallography	2	Criminology & Penology	5
Dermatology & Venereal Diseases	2	Family Studies	5
Hematology	2	Psychology, Applied	5
Environmental Sciences	2	Philosophy	5
Clinical Neurology	2	Communication	5
Pathology	2	Literature, American	5
Obstetrics & Gynecology	2	Art	5

Table 7.47. (Cont.)

Category	Distance	Category	Distance
Ophthalmology	2	Materials Science, Multidisciplinary	6
Pediatrics	2	Mathematics, Applied	6
Polymer Science	2	Physics, Applied	6
Engineering, Chemical	2	Engineering, Mechanical	6
Parasitology	2	History & Philosophy Of Science	6
Virology	2	Engineering, Marine	6
Substance Abuse	2	Computer Science, Interdisciplinary Applications	6
Chemistry, Organic	2	Imaging Science & Photographic Technology	6
Chemistry, Physical	2	Women's Studies	6
Chemistry, Inorganic & Nuclear	2	Language & Linguistics	6
Behavioral Sciences	2	Poetry	6
Toxicology	2	Architecture	6
Geriatrics & Gerontology	2	Mathematics	7
Forestry	2	Medical Informatics	7
Infectious Diseases	2	Construction & Building Technology	7
Rheumatology	2	Metallurgy & Metallurgical Engineering	7
Urology & Nephrology	2	Materials Science, Composites	7
Allergy	2	Engineering, Electrical & Electronic	7
Biology, Miscellaneous	2	Engineering	7
Public, Environmental & Occupational Health	2	Materials Science, Ceramics	7
Emergency Medicine & Critical Care	2	Engineering, Biomedical	7
Rehabilitation	2	Operations Research & Management Science	7
Sport Sciences	2	Materials Science, Coatings & Films	7
Andrology	2	Mathematics, Miscellaneous	7
Spectroscopy	2	Materials Science, Characterization & Testing	7
Horticulture	2	Literature, Romance	7
Transplantation	2	Literature	7
Electrochemistry	2	Computer Science, Theory & Methods	8
Mycology	2	Statistics & Probability	8
Microscopy	2	Automation & Control Systems	8

Table 7.47. (Cont.)

Category	Distance	Category	Distance
Social Sciences, Interdisciplinary	2	Computer Science, Artificial Intelligence	8
Social Issues	2	Telecommunications	8
Nursing	2	Computer Science, Hardware & Architecture	8
Health Care Sciences & Services	2	Materials Science, Biomaterials	8
Music	2	Computer Science, Cybernetics	8
Energy & Fuels	3	Engineering, Industrial	8
Geosciences, Interdisciplinary	3	Transportation	8
Acoustics	3	Social Sciences, Mathematical Methods	8
Respiratory System	3	Management	8
Water Resources	3	History	8
Dentistry, Oral Surgery & Medicine	3	Theater	8
Ecology	3	Literary Reviews	8
Orthopedics	3	Folklore	8
Otorhinolaryngology	3	Literature, British Isles	8
Psychiatry	3	Computer Science, Information Systems	9
Tropical Medicine	3	Computer Science, Software, Graphics, Programming	9
Physics, Atomic, Molecular & Chemical	3	Engineering, Manufacturing	9
Medicine, Legal	3	Ergonomics	9
Anthropology	3	Economics	9
Peripheral Vascular Disease	3	Business	9
Instruments & Instrumentation	3	History Of Social Sciences	9
Engineering, Environmental	3	Ethnic Studies	9
Chemistry, Applied	3	Religion	9
Materials Science, Paper & Wood	3	Classics	9
Thermodynamics	3	Literature, African, Australian, Canadian	9
Environmental Studies	3	Agricultural Economics & Policy	10
Social Sciences, Biomedical	3	Information Science & Library Science	10
Psychology, Biological	3	Law	10
Education, Special	3	Planning & Development	10

Table 7.47. (Cont.)

Category	Distance	Category	Distance
Health Policy & Services	3	Business, Finance	10
Demography	3	Public Administration	10
Arts & Humanities, General	3	Area Studies	10
Literature, German, Netherlandic, Scandinavian	3	Political Science	10
Engineering, Petroleum	4	International Relations	10
Astronomy & Astrophysics	4	Industrial Relations & Labor	10
Marine & Freshwater Biology	4	Asian Studies	10
Physics, Multidisciplinary	4		

Table 7.48. Distances of the categories of the scientogram of Spain's scientific output, 1999–2002, with respect to the central category

Category	Distance	Category	Distance
Biochemistry & Molecular Biology	0	Ornithology	4
Medicine, General & Internal	1	Environmental Studies	4
Chemistry, Multidisciplinary	1	Psychology, Clinical	4
Agriculture	1	Materials Science, Textiles	4
Nutrition & Dietetics	1	Archeology	4
Plant Sciences	1	Psychology, Psychoanalysis	4
Zoology	1	Demography	4
Biology	1	Literary Reviews	4
Biotechnology & Applied Microbiology	1	Biodiversity Conservation	4
Biochemical Research Methods	1	Literary Theory & Criticism	4
Cell Biology	1	Engineering, Petroleum	5
Endocrinology & Metabolism	1	Mechanics	5
Medicine, Research & Experimental	1	Physics, Nuclear	5
Neurosciences	1	Optics	5
Oncology	1	Physics, Condensed Matter	5
Pharmacology & Pharmacy	1	Oceanography	5
Physiology	1	Nuclear Science & Technology	5
Veterinary Sciences	1	Physics, Fluids & Plasmas	5
Microbiology	1	Physics, Particles & Fields	5
Biophysics	1	Fisheries	5
Medical Laboratory Technology	1	Limnology	5

Table 7.48. (Cont.)

Category	Distance	Category	Distance
Reproductive Systems	1	Psychology, Experimental	5
Genetics & Heredity	1	Psychology, Mathematical	5
Immunology	1	Engineering, Civil	5
Developmental Biology	1	Physics, Mathematical	5
Chemistry, Medicinal	1	Psychology, Developmental	5
Film, Radio, Television	1	Education & Educational Research	5
Gastroenterology & Hepatology	2	Psychology, Educational	5
Radiology, Nuclear Medicine & Medical Imaging	2	Sociology	5
Education, Scientific Disciplines	2	Urban Studies	5
Entomology	2	Social Work	5
Cardiac & Cardiovascular Systems	2	Psychology, Social	5
Agriculture, Dairy & Animal Science	2	Criminology & Penology	5
Agriculture, Soil Science	2	Family Studies	5
Food Science & Technology	2	Psychology, Applied	5
Anesthesiology	2	Communication	5
Anatomy & Morphology	2	Architecture	5
Surgery	2	Literature, African, Australian, Canadian	5
Crystallography	2	Psychology, Multidisciplinary	5
Dermatology & Venereal Diseases	2	Geosciences, Interdisciplinary	6
Hematology	2	Materials Science, Multidisciplinary	6
Clinical Neurology	2	Mathematics, Applied	6
Pathology	2	Physics, Applied	6
Obstetrics & Gynecology	2	Engineering, Mechanical	6
Ophthalmology	2	Engineering, Marine	6
Pediatrics	2	Computer Science, Interdisciplinary Applications	6
Parasitology	2	Language & Linguistics	6
Virology	2	Ethnic Studies	6
Chemistry, Organic	2	Mathematics	7
Chemistry, Analytical	2	Medical Informatics	7
Chemistry, Physical	2	Construction & Building Technology	7
Chemistry, Inorganic & Nuclear	2	Computer Science, Software, Graphics, Programming	7
Behavioral Sciences	2	Astronomy & Astrophysics	7

Table 7.48. (Cont.)

Category	Distance	Category	Distance
Toxicology	2	Metallurgy & Metallurgical Engineering	7
Geriatrics & Gerontology	2	Statistics & Probability	7
Forestry	2	Materials Science, Composites	7
Infectious Diseases	2	Engineering, Electrical & Electronic	7
Rheumatology	2	Meteorology & Atmospheric Sciences	7
Urology & Nephrology	2	Engineering	7
Allergy	2	Engineering, Manufacturing	7
Biology, Miscellaneous	2	Paleontology	7
Public, Environmental & Occupational Health	2	Materials Science, Ceramics	7
Emergency Medicine & Critical Care	2	Geochemistry & Geophysics	7
Rehabilitation	2	Engineering, Biomedical	7
Sport Sciences	2	Operations Research & Management Science	7
Andrology	2	Materials Science, Coatings & Films	7
Horticulture	2	Geology	7
Transplantation	2	Materials Science, Characterization & Testing	7
Mycology	2	Mining & Mineral Processing	7
Microscopy	2	Engineering, Geological	7
Social Issues	2	Literature, Romance	7
Health Care Sciences & Services	2	Engineering, Ocean	7
Poetry	2	Music	7
Folklore	2	Literature	7
Agriculture, Multidisciplinary	2	Applied Linguistics	7
Integrative & Complementary Medicine	2	Geography, Physical	7
Medical Ethics	2	Computer Science, Theory & Methods	8
Agricultural Engineering	2	Computer Science, Information Systems	8
Acoustics	3	Engineering, Aerospace	8
Respiratory System	3	Automation & Control Systems	8
Environmental Sciences	3	Computer Science, Artificial Intelligence	8
Dentistry, Oral Surgery & Medicine	3	Telecommunications	8

Table 7.48. (Cont.)

Category	Distance	Category	Distance
Ecology	3	Mineralogy	8
Orthopedics	3	Computer Science, Hardware & Architecture	8
Otorhinolaryngology	3	Mathematics, Miscellaneous	8
Polymer Science	3	Geography	8
Engineering, Chemical	3	Materials Science, Biomaterials	8
Psychiatry	3	Computer Science, Cybernetics	8
Tropical Medicine	3	Engineering, Industrial	8
Physics, Atomic, Molecular & Chemical	3	Remote Sensing	8
Medicine, Legal	3	Management	8
Anthropology	3	Social Sciences, Interdisciplinary	8
Peripheral Vascular Disease	3	Philosophy	8
Chemistry, Applied	3	Theater	8
Materials Science, Paper & Wood	3	Literature, British Isles	8
Spectroscopy	3	Transportation Science & Technology	8
Electrochemistry	3	History & Philosophy Of Science	9
Social Sciences, Biomedical	3	Information Science & Library Science	9
Psychology, Biological	3	Ergonomics	9
Nursing	3	Transportation	9
Women's Studies	3	Imaging Science & Photographic Technology	9
Education, Special	3	Social Sciences, Mathematical Methods	9
Health Policy & Services	3	Business	9
Literature, American	3	Planning & Development	9
Arts & Humanities, General	3	Robotics	9
Art	3	Ethics	9
Literature, German, Netherlandic, Scandinavian	3	Economics	10
Critical Care Medicine	3	Agricultural Economics & Policy	11
Gerontology	3	Law	11
Neuroimaging	3	Business, Finance	11
Evolutionary Biology	3	Public Administration	11
Energy & Fuels	4	Area Studies	11
Water Resources	4	Political Science	11
Marine & Freshwater Biology	4	International Relations	11
Physics, Multidisciplinary	4	History Of Social Sciences	11

Table 7.48. (Cont.)

Category	Distance	Category	Distance
Substance Abuse	4	Industrial Relations & Labor	11
Instruments & Instrumentation	4	History	12
Engineering, Environmental	4	Religion	13
Psychology	4	Classics	13
Thermodynamics	4	Asian Studies	14

On a general level, the degree of universality of the categories is very similar over the three periods. The main differences are found above all in the area of *Humanities* which, because of its sometimes dubious relations, presents great variations in its distances. As for the rest, the degree to which they share sources with other categories or participate in the development of the domain is perfectly defined by the geodesic distance in each period.

For instance, if we wish to find the 20 categories that participate most in research over the three time periods, we only need to select those with the least distance in each interval (Table 7.49).

Table 7.49. Twenty categories of the most universal character in the Spanish scientific domain, 1990–2002

Category	1990–1994 distance	1995–1998 distance	1999–2002 distance
Biochemistry & Molecular Biology	0	0	0
Agricultura	1	1	1
Biochemical Research Methods	1	1	1
Biology	1	1	1
Biophysics	1	1	1
Biotechnology & Applied Microbiology	1	1	1
Chemistry, Medicinal	1	1	1
Chemistry, Multidisciplinary	1	1	1
Developmental Biology	1	1	1
Endocrinology & Metabolism	1	1	1
Genetics & Heredity	1	1	1
Immunology	1	1	1
Medicine, Research & Experimental	1	1	1
Microbiology	1	1	1
Neurosciences	1	1	1
Nutrition & Dietetics	1	1	1
Oncology	1	1	1
Pharmacology & Pharmacy	1	1	1
Physiology	1	1	1
Plant Sciences	1	1	1
Reproductive Systems	1	1	1
Veterinary Sciences	1	1	1
Zoology	1	1	1

Prominence

As seen in either tabular or graphic form, *Biochemistry & Molecular Biology* is the most prominent category of the three periods, and its degree of prominence increases with the passing of time (Table 7.50).

Table 7.50. Most prominent categories of the Spanish domain, 1990–2002

1990–1994 category	Prominence	1995–1998 category	Prominence
Biochemistry & Molecular Biology	151.53	Biochemistry & Molecular Biology	187.68
Medicine, General & Internal	52.58	Psychology	38.6
Psychology	38.77	Geosciences, Interdisciplinary	37.29
Physics, Multidisciplinary	25.82	Mathematics, Applied	26.46
Engineering, Electrical & Electronic	22.4	Engineering, Electrical & Electronic	25
Geosciences, Interdisciplinary	21.88	Physics, Multidisciplinary	23.3
Mathematics, Applied	19.6	Economics	20.54
Language & Linguistics	15.53	History	15
History	15	Literature, Romance	12.4
Economics	14.4	Materials Science, Multidisciplinary	12.2
Materials Science, Multidisciplinary	10.29	Biology, Miscellaneous	11.53
Marine & Freshwater Biology	9	Marine & Freshwater Biology	11

1999–2002 category	Prominence
Biochemistry & Molecular Biology	201.75
Psychology	37.75
Geosciences, Interdisciplinary	29
Engineering, Electrical & Electronic	26.5
Physics, Multidisciplinary	21.82
Economics	21.54
Biology, Miscellaneous	21.32
Chemistry, Analytical	20.32
Materials Science, Multidisciplinary	18.75
Mathematics, Applied	18.7125
Language & Linguistics	11.47

For the period 1990–1994, the robbery algorithm detects 23 prominent categories (Fig. 7.29). By means of the scree test we identify 12, 9 of which are the origin of 1 of the 10 thematic areas of this period. What remains is to identify the area *Agriculture and Soil Science*, whose category of origin we hold to be *Agriculture*.

In the interval 1995–1998, the robbery algorithm detects 20 categories (Fig. 7.30). Twelve are identified by scree test, and nine of these give rise to the ten thematic areas of this period. *Agriculture and Soil Sciences* is not identified as prominent. Again, we deem its source category to be *Agriculture*.

Finally, in the period 1999–2002, 20 prominent categories are identified, 11 of them likewise identified by the scree test (Fig. 7.31). Nine of these are the origin of one of the eleven thematic areas. Undetected go *Health Policy and Medical Services* and *Agriculture and Soil Sciences*, whose source categories we consider to be *Agriculture* and *Medicine General & Internal*.

To sum up, and in view of the results obtained using the robbery algorithm of degree, we can state that *Biochemistry & Molecular Biology* is consolidated as the most prominent category of the Spanish scientific domain in terms of output and visibility, and it is on the rise, the utilization of its sources increasingly referred to by the surrounding categories as time passes. *Psychology*, after the second period and up to 2002, has come forth as the second most prominent category of Spain's domain, although its degree of prominence does not increase during the final period, but remains stable. Finally, *Geosciences* is the third category to stand out. It takes hold of this position after the second period of study, and rivals hard with *Engineering Electrical & Electronics—Computer Science and Telecommunications*—for the bronze medal.

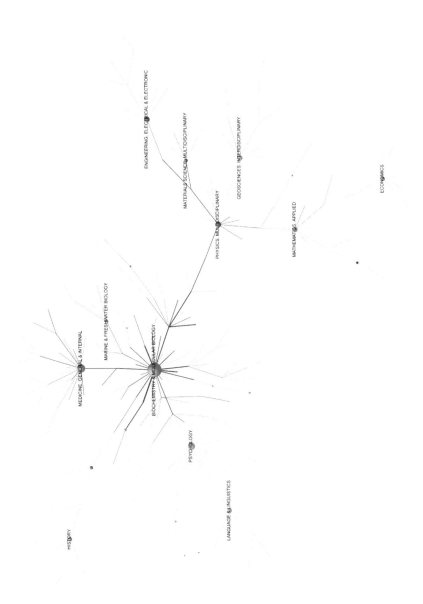

Fig. 7.29. Scientogram of the most prominent categories of the Spanish domain, 1990–1994

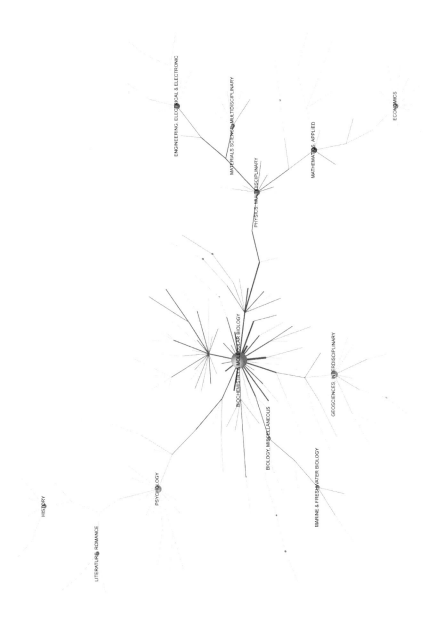

Fig. 7.30. Scientogram of the most prominent categories of the Spanish domain, 1995–1998

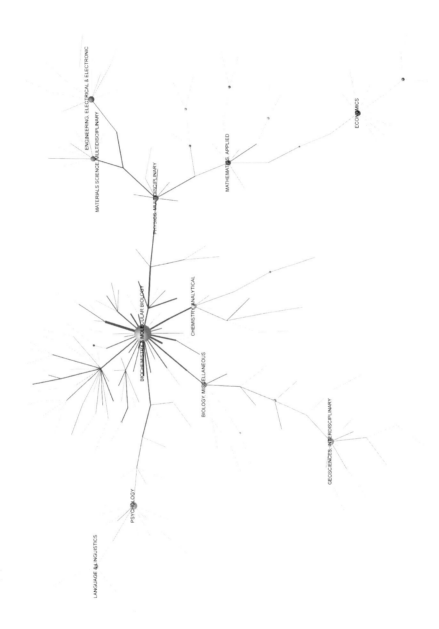

Fig. 7.31. Scientogram of the most prominent categories of the Spanish domain, 1999–2002

7.3.4 Backbone

Evolution, in the context of our research, tells us that the Backbone of Spain's scientific research is growing stronger over the years.

In the period 1990–1994, there are three thematic areas that make up the backbone of Spanish science: *Biomedicine, Materials Science & Applied Physics,* and *Agriculture and Soil Sciences,* which comprise seven categories (Fig. 7.32). Yet three of these red categories come to represent what what we referred to in Sect. 7.1.3.2. as "type one" categories; that is, those mirroring the early first evolutive stage of a thematic area.

In the interval 1995–1998, the situation is very similar, and the backbone of research consists of two consolidated thematic areas plus one emerging one (Fig. 7.33). But there is an increase in the number of categories in *Biomedicine* (three) as well as a reduction to just one in the area of *Agriculture and Soil Sciences.* So the total number of categories rises with respect to the previous period – from seven to ten.

In the final period, three thematic areas make up the backbone of Spanish research (seeing as *Agriculture and Soil Sciences* is consolidated in the meantime, as shown in Fig. 7.34). The number of categories that make up these three areas also grows with respect to the previous period: from 10 to 13. These findings lead us to confirm that Spain's presence is solidified over time.

On a general level, and in view of Figs. 7.32–7.34, we can only reaffirm what we have stated thus far on a macro- and microstructural level:

- The thematic area of *Biomedicine* is the nucleus of research in Spain's scientific domain; it is the area sharing more resources with others and

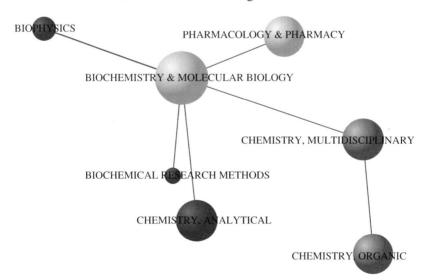

Fig. 7.32. Backbone of Spanish research 1990–1994

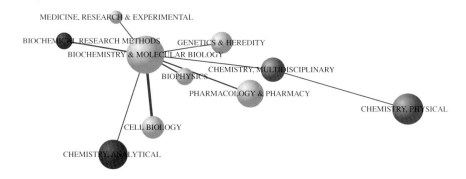

Fig. 7.33. Backbone of Spain's scientific research, 1995–1998

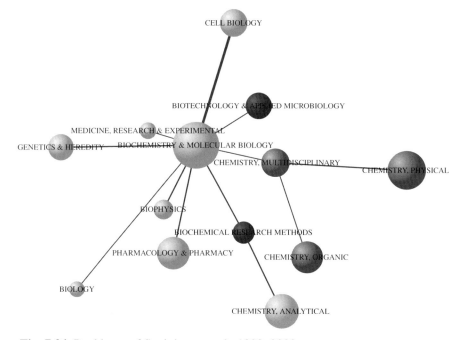

Fig. 7.34. Backbone of Spain's research, 1999–2002

the one to participate most in the development of science. Over time, it is also the one to increase its categories, and therefore its influence over other ones, the most.

- The second most influential thematic area of the Spanish domain is *Materials Science and Applied Physics*, which similarly evolves and develops over the years, though to a lesser degree.

A third noteworthy thematic area is *Agriculture and Soil Sciences*. However, further measures such as prominence indicate that this third position could likewise be claimed by *Earth and Space Sciences*, or even, by virtue of the centrality of degree on the macrostructural level, by *Biology Animal and Ecology*. These divergences in lower positions are common with some of the measures based on the degree of social networks. What is important is that with only slight variations they coincide in pointing out which are the most significant areas.

8 Discussion

Scientography of vast scientific domains based on the raw cocitation of JCR categories and represented using PFNET lends us the possibility of exploring the state of investigation from a wide array of perspectives. On the one hand, it offers domain analysts the capacity to discover the most significant connections between categories of a given domain. It also allows one to see how these categories are grouped into major thematic areas and how they are interrelated through a logical internal. Finally, it facilitates the detailed analysis of the structure of a domain, its comparison with another, or the study of its evolution over time.

Meanwhile, those users interested in information retrieval may access the documents found behind each one of the spheres that represent the categories of the scientograms, and they can access as well the works contained in the links between categories, where the intellectual substance is interchanged, to thereby appreciate the influence of one category on another. Moreover, it is a fairly simple matter to make accessible the full text of those studies from institutions, such as the University of Granada, where users already have electronic access to the contents of the worlds main scientific journals.

Lastly, scientograms offer new researchers and new viewers a lasting image of the essential structure of a domain, which can help complete the mental image one may have begun to formulate, or become a point of reference from which to begin perceiving a vast scientific domain.

8.1 Importance and Quality of Results

The quality and significance of the results of our scientograms can be evaluated from at least three separate perspectives.

1. From the point of view of cocitation and information visualization, as we saw throughout Chap. 7.
2. From the comparative consideration of a thematic classification accepted as an evaluative element of science, as we shall see in the following.

3. Through user-based evaluation. This perspective should not be underestimated, as, given the long-reaching relevance of such a task, it is held up as a future line of research.

There are a number of thematic classifications for grouping JCR categories into superior conglomerations. Among these is that of the Spanish *Agencia Nacional de Evaluación y Prosepectiva* (ANEP, 2005), which we chose because it is the one used for the technical assessment and scientific evaluation within the National Plan for Spanish Research and Development (*Plan Nacional de Investigación y Desarrolllo Español*). The taxonomy taken up by the ANEP groups the JCR categories into 25 major classes – Annex V – of which we eliminated the "multidisiciplinary" category. The evaluation of our scientograms consists, then, in substituting the name of the JCR category with that of the ANEP class. The generosity of the classification system of the scientograms, embodying the combination of PFNET plus raw category cocitation, will be greater or lesser depending on the extent that the thematic areas of the ANEP coincide with the bunchings of the scientograms. At the same time, the utility of factor scientograms will remain manifest insofar as these groupings coincide with the thematic areas indicated by FA. The ANEP classification allows for the multiple assignment of a single category to different classes. Because in the scientograms a single category can belong to only one thematic area, in solving this problem of ascription of a category to one class or another we chose the first assignment made by the ANEP and rejected the rest. In order to get a grasp, with a quick view, of the degree of fit of the taxonomy proposed with our scientograms as opposed to that established by the ANEP, we have used the abbreviations of their classes, whose equivalency is given in Annex V.

Below we show the abbreviations of the ANEP classes overlaying the evolutive factor scientograms of the Spanish scientific domain over the three periods of study (Figs. 8.1–8.3) .

If we observe closely any one of these three evolutive scientograms (Figs. 8.1–8.3), we notice that the coincidence between category groupings (ours vs. those of the ANEP) is practically complete. The discrepancies existing owe mainly to the difference of the number of factors identified by FA, and to the number of classes that, according to the ANEP, integrate Spain's scientific domain. For instance, in the factor scientogram of classes for the period 1999–2002, the first factor, which we call *Biomedicine*, almost totally coincides with those two the ANEP considers to make up this thematic area: Biology Molecular Cellular and Genetics (MOL) and Medicine (MED). The same can be said of the second factor identified, *Materials Sciences and Applied Physics*, and its ANEP counterparts: Science and Technology of Materials (MAR), *Physics and Space Sciences* (FIS), and

Chemistry (QUI). We could go on to point out *Psychology* and its correspondence with Psychology and Education Sciences (PSI), or *Earth and Space Sciences* with their counterpart Earth Sciences (TIE), or *Computers Science and Telecommunications* with Computation Sciences and Information Technology (COM), and many more. All 11 factors identified in this period could be evoked to ratify the equivalency and therefore the confirmed quality of the results of our scientograms.

The divergences produced between the groupings proposed by the scientograms and by the ANEP classification must be considered as singularities that contribute with additional information about the specific characteristics of the domains. We must recall that the relations revealed in the scientograms are no more than the reflection of the unconscious labor of hundreds of thousands of researchers coming to the surface through their citations. In contrast, the classification of the ANEP classification is a taxonomy elaborated by a handful of experts.

Fig. 8.1. Factor scientogram of ANEP classes of the Spanish domain, 1990–1994

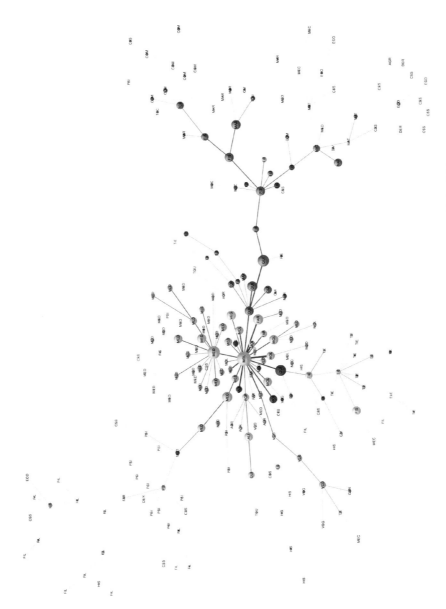

Fig. 8.2. Factor scientogram of ANEP classes of the Spanish domain, 1995–1998

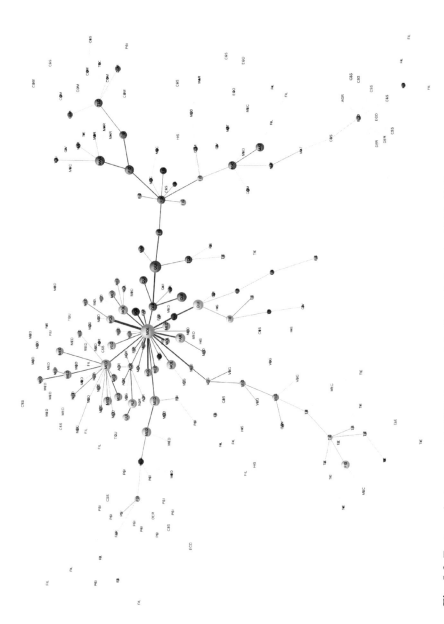

Fig. 8.3. Factor scientogram of ANEP classes of the Spanish domain, 1999–2002

9 Epilogue

The informational end products deriving from this work, as expounded over the preceding eight chapters, stand as the visual and logical consequence of the diverse methodologies of representation and analysis we have applied and advocated. Still, some general conclusions deserve space as well, alongside the outlines of our future research.

The baptisms of fire of this methodology for the visualization and analysis of major scientific domains, applied to diverse geographic settings, are recorded and displayed here. The far-reaching capabilities of current techniques and the desire to explore vast horizons led us to embark on creating a map of the greatest geographic domain possible in the intellectual universe: the world. Yet the world of scientific knowledge and progress can also be broken down into extensive constituents, such as the United States and the European Union, as we have seen. And the search for advancement, requiring curiosity, determination, intercommunication, and "daring to know," also urged us to closely examine our own role, through the visualization and evolutive study of the scientific domain of Spain itself.

In full view of the fact that the scientograms of these geographic domains are conceived using the information contained in ISI databases, the application of the proposed methodology to other types of domains – whether they be thematic, institutional, or of another nature – should present no imposing difficulties.

This is, indeed, a powerful tool. Its potential is so great and ample that we chose to document its conception. This turning point in information visualization makes it possible to take in and analyze the complete groundworks or skeletal system – the macrostructure and the microstructure – of a vast domain of knowledge and communication. We insist that this can be done, as in our case, with basic media and minimal informational costs, even for a domain of millions of documents and the relations they entail.

The power and potential of this tool go so far as to depict relational information chained together by a series of intelligible sequences, which can be seen as the most basic elements of comprehension, analysis, and interpretation, and therefore the most adequate way to represent concentrations of information and the flux of knowledge. Understanding is by no means the exclusive realm of experts, as these user-friendly depictions prove.

Yet scientograms should not be considered the simple graphic representation of the JCR structure. They transcend to stand as proof of the evolution of science. In this sense they may also be interpreted as the consensual opinion of scientific researchers (or other participants of a given domain) as to the collective image they hold of themselves. We must acknowledge at the same time, however, that these scientograms can be improved and further clarified by document categorization directly from author citation in association with JCR categories; that is, bypassing the journals, as we did in this case with multidisciplinary documents. The time required for such a categorization of documents is considerable, and will have to be relegated to future research attempts.

With a surprisingly low informational investment, and as a clustering method, it is perfectly feasible to superpose even greater structures upon the foundations of imagery laid down here. Such is the case of the macro-structure obtained using FA, which demonstrates the high degree of compatibility and complementarity of the two procedures. Other structures of a hierarchically superior nature, such as the ANEP classification or ISI fields, could likewise be used as the agglutinating element that would double as the evaluative parameter of the structure.

Information retrieval through categories and links facilitates user appraisal of the meaning and utility of scientograms, as the most accessible reflection of the complexity and evolution of a domain. The practical application of this methodology represents the convergence of different fields of investigation, including information visualization, citation analysis, social network analysis, and domain analysis.

Scientograms cannot predict the future of science, but they can indicate where we are, and whence we have evolved. The changes that will become apparent through future scientograms, year after year, will no doubt reveal trends that can be used for the diagnosis and prognosis of the state of that domain.

A geographical grouping turns these visualizations into a historical atlas of sorts, representing scientific ideas and advancements over a specific time period in which the relations between and among disciplines, or between more particular thematic areas, come to highlight the main focal points of research and interaction, and the predominating lines.

We believe our scientograms encompass an unfailing design, which lends them for use as tools for widespread visual–analytical applications, because of the following properties:

- They represent both small and large amounts of information.
- They reduce the time of visual search.
- They facilitate comprehension of complex data structures.

- They make manifest relations between and among elements that would otherwise go unacknowledged.
- They encourage deeper thought and the formulation of hypotheses, and
- They stand to become the object of analysis, interest, and discussion.

There are multiple advantages to having scientograms for major domains, in terms of information visualization and its analysis. Yet these displays might be improved on two fronts. First, with regard to the information used to construct the visualizations: although the ISI databases are to date one of the most prestigious and most appropriate resources for drawing forth the scientific structure of any domain, the exhaustivity of scientograms would be enhanced by the incorporation of information from other sources, such as databases from specialized fields or conference reports, among other possible sources.

Finally, the interpretation of scientograms deserves mention. Although they are the objective reflection of a domain, representing as they do the consensual opinion of thousands of authors who selectively cite their most relevant sources, it is possible that our interpretation and processing of these citations entails some degree of individual subjectivity. Whether knowledge can be depicted or grasped without the intervention of human subjectivity is a matter that cannot be resolved, but should be brought to light.

Annex I

Countries Producing Scientific Documents, According to the ISI

1	Afghanistan	26	Bulgaria	51	Egypt	76	Haiti
2	Albania	27	Burkina Faso	52	El Salvador	77	Honduras
3	Algeria	28	Burundi	53	England	78	Hungary
4	Andorra	29	Byelarus	54	Equat Guinea	79	Iceland
5	Angola	30	Cambodia	55	Eritrea	80	India
6	Antigua & Barbu	31	Cameroon	56	Estonia	81	Indonesia
7	Argentina	32	Canada	57	Ethiopia	82	Iran
8	Armenia	33	Cent Afr Republ	58	Fiji	83	Iraq
9	Australia	34	Chad	59	Finland	84	Ireland
10	Austria	35	Chile	60	Fr Polynesia	85	Israel
11	Azerbaijan	36	Colombia	61	France	86	Italy
12	Bahamas	37	Comoros	62	French Guiana	87	Jamaica
13	Bahrain	38	Congo	63	Gabon	88	Japan
14	Bangladesh	39	Cook Islands	64	Gambia	89	Jordan
15	Barbados	40	Costa Rica	65	Germany	90	Kazakhstan
16	Belgium	41	Cote Ivoire	66	Ghana	91	Kenya
17	Belize	42	Croatia	67	Gilbraltar	92	Kuwait
18	Benin	43	Cuba	68	Greece	93	Kyrgyzstan
19	Bermuda	44	Cyprus	69	Greenland	94	Laos
20	Bhutan	45	Czech Republic	70	Grenada	95	Latvia
21	Bolivia	46	Denmark	71	Guadeloupe	96	Lebanon
22	Bosnia & Herceg	47	Djibouti	72	Guatemala	97	Lesotho
23	Botswana	48	Dominica	73	Guinea	98	Liberia
24	Brazil	49	Dominican Rep	74	Guinea Bissau	99	Libya
25	Brunei	50	Ecuador	75	Guyana	100	Liechtenstein

101	Lithuania	128	New Caledonia	155	Scotland	182	Togo
102	Luxembourg	129	New Zealand	156	Senegal	183	Tonga
103	Macao	130	Nicaragua	157	Seychelles	184	Trinid & Tobago
104	Macedonia	131	Niger	158	Sierra Leone	185	Tunisia
105	Madagascar	132	Nigeria	159	Singapore	186	Turkey
106	Malagasy Republ	133	North Ireland	160	Slovakia	187	Turkmenistan
107	Malawi	134	North Korea	161	Slovenia	188	U Arab Emirates
108	Malaysia	135	Norway	162	Solomon Islands	189	Uganda
109	Maldives	136	Oman	163	Somalia	190	Ukraine
110	Mali	137	Pakistan	164	South Africa	191	Uruguay
111	Malta	138	Palau	165	South Korea	192	USA
112	Marshall Island	139	Panama	166	Spain	193	Uzbekistan
113	Martinique	140	Papua N Guinea	167	Sri Lanka	194	Vanuatu
114	Mauritania	141	Paraguay	168	St Kitts & Nevi	195	Vatican
115	Mauritius	142	Peoples R China	169	St Lucia	196	Venda
116	Mexico	143	Peru	170	St Vincent	197	Venezuela
117	Micronesia	144	Philippines	171	Sudan	198	Vietnam
118	Moldova	145	Poland	172	Surinam	199	W Ind Assoc St
119	Monaco	146	Portugal	173	Swaziland	200	Wales
120	Mongol Peo Rep	147	Qatar	174	Sweden	201	Western Samoa
121	Morocco	148	Rep of Georgia	175	Switzerland	202	Yemen
122	Mozambique	149	Reunion	176	Syria	203	Yugoslavia
123	Myanmar	150	Romania	177	Taiwan	204	Zaire
124	Namibia	151	Russia	178	Tajikistan	205	Zambia
125	Nepal	152	Rwanda	179	Tajikstan	206	Zimbabwe
126	Neth Antilles	153	Sao Tome & Prin	180	Tanzania		
127	Netherlands	154	Saudi Arabia	181	Thailand		

JCR Categories 2002

1	Acoustics	31	Chemistry, Inorganic & Nuclear
2	Agricultural Economics & Policy	32	Chemistry, Medicinal
3	Agriculture	33	Chemistry, Multidisciplinary
4	Agriculture, Dairy & Animal Science	34	Chemistry, Organic
5	Agriculture, Soil Science	35	Chemistry, Physical
6	Allergy	36	Clinical Neurology
7	Anatomy & Morphology	37	Communication
8	Andrology	38	Computer Science, Artificial Intelligence
9	Anesthesiology	39	Computer Science, Cybernetics
10	Anthropology	40	Computer Science, Hardware & Architecture
11	Archeology	41	Computer Science, Information Systems
12	Architecture	42	Computer Science, Interdisciplinary Applications
13	Area Studies	43	Computer Science, Software, Graphics, Programming
14	Art	44	Computer Science, Theory & Methods
15	Arts & Humanities, General	45	Construction & Building Technology
16	Astronomy & Astrophysics	46	Criminology & Penology
17	Automation & Control Systems	47	Crystallography
18	Behavioral Sciences	48	Demography
19	Biochemical Research Methods	49	Dentistry, Oral Surgery & Medicine
20	Biochemistry & Molecular Biology	50	Dermatology & Venereal Diseases
21	Biology	51	Developmental Biology
22	Biology, Miscellaneous	52	Ecology
23	Biophysics	53	Economics
24	Biotechnology & Applied Microbiology	54	Education & Educational Research
25	Business	55	Education, Scientific Disciplines
26	Business, Finance	56	Education, Special
27	Cardiac & Cardiovascular Systems	57	Electrochemistry
28	Cell Biology	58	Emergency Medicine & Critical Care
29	Chemistry, Analytical	59	Endocrinology & Metabolism
30	Chemistry, Applied	60	Energy & Fuels

61	Engineering	96	History Of Social Sciences
62	Engineering, Aerospace	97	Horticulture
63	Engineering, Biomedical	98	Imaging Science & Photographic Technology
64	Engineering, Chemical	99	Immunology
65	Engineering, Civil	100	Industrial Relations & Labor
66	Engineering, Electrical & Electronic	101	Infectious Diseases
67	Engineering, Environmental	102	Information Science & Library Science
68	Engineering, Geological	103	Instruments & Instrumentation
69	Engineering, Industrial	104	International Relations
70	Engineering, Manufacturing	105	Language & Linguistics
71	Engineering, Marine	106	Law
72	Engineering, Mechanical	107	Limnology
73	Engineering, Ocean	108	Literary Reviews
74	Engineering, Petroleum	109	Literature
75	Entomology	110	Literature, American
76	Environmental Sciences	111	Literature, Romance
77	Environmental Studies	112	Literature, Slavic
78	Ergonomics	113	Management
79	Ethnic Studies	114	Marine & Freshwater Biology
80	Family Studies	115	Materials Science, Biomaterials
81	Fisheries	116	Materials Science, Ceramics
82	Food Science & Technology	117	Materials Science, Characterization & Testing
83	Forestry	118	Materials Science, Coatings & Films
84	Gastroenterology & Hepatology	119	Materials Science, Composites
85	Genetics & Heredity	120	Materials Science, Multidisciplinary
86	Geochemistry & Geophysics	121	Materials Science, Paper & Wood
87	Geography	122	Materials Science, Textiles
88	Geology	123	Mathematics
89	Geosciences, Interdisciplinary	124	Mathematics, Applied
90	Geriatrics & Gerontology	125	Mathematics, Miscellaneous
91	Health Care Sciences & Services	126	Mechanics
92	Health Policy & Services	127	Medical Informatics
93	Hematology	128	Medical Laboratory Technology
94	History	129	Medicine, General & Internal
95	History & Philosophy of Science	130	Medicine, Legal

131	Medicine, Research & Experimental	166	Physics, Multidisciplinary
132	Metallurgy & Metallurgical Engineering	167	Physics, Nuclear
133	Meteorology & Atmospheric Sciences	168	Physics, Particles & Fields
134	Microbiology	169	Physiology
135	Microscopy	170	Planning & Development
136	Mineralogy	171	Plant Sciences
137	Mining & Mineral Processing	172	Poetry
138	Multidisciplinary Sciences	173	Political Science
139	Music	174	Polymer Science
140	Mycology	175	Psychiatry
141	Neurosciences	176	Psychology
142	Nuclear Science & Technology	177	Psychology, Applied
143	Nursing	178	Psychology, Biological
144	Nutrition & Dietetics	179	Psychology, Clinical
145	Obstetrics & Gynecology	180	Psychology, Developmental
146	Oceanography	181	Psychology, Educational
147	Oncology	182	Psychology, Experimental
148	Operations Research & Management Science	183	Psychology, Mathematical
149	Ophthalmology	184	Psychology, Psychoanalysis
150	Optics	185	Psychology, Social
151	Ornithology	186	Public Administration
152	Orthopedics	187	Public, Environmental & Occupational Health
153	Otorhinolaryngology	188	Radiology, Nuclear Medicine & Medical Imaging
154	Paleontology	189	Rehabilitation
155	Parasitology	190	Religion
156	Pathology	191	Remote Sensing
157	Pediatrics	192	Reproductive Biology
158	Peripheral Vascular Disease	193	Respiratory System
159	Pharmacology & Pharmacy	194	Rheumatology
160	Philosophy	195	Social Issues
161	Physics, Applied	196	Social Sciences, Biomedical
162	Physics, Atomic, Molecular & Chemical	197	Social Sciences, Interdisciplinary
163	Physics, Condensed Matter	198	Social Sciences, Mathematical Methods
164	Physics, Fluids & Plasmas	199	Social Work
165	Physics, Mathematical	200	Sociology

201	Spectroscopy	211	Transportation
202	Sport Sciences	212	Tropical Medicine
203	Statistics & Probability	213	Urban Studies
204	Substance Abuse	214	Urology & Nephrology
205	Surgery	215	Veterinary Sciences
206	Telecommunications	216	Virology
207	Theater	217	Water Resources
208	Thermodynamics	218	Women's Studies
209	Toxicology	219	Zoology
210	Transplantation		

JCR Categories 1990–2002

1	Acoustics	26	Biology
2	Agricultural Economics & Policy	27	Biology, Miscellaneous
3	Agricultural Engineering	28	Biophysics
4	Agriculture	29	Biotechnology & Applied Microbiology
5	Agriculture, Dairy & Animal Science	30	Business
6	Agriculture, Multidisciplinary	31	Business, Finance
7	Agriculture, Soil Science	32	Cardiac & Cardiovascular Systems
8	Allergy	33	Cell Biology
9	Anatomy & Morphology	34	Chemistry, Analytical
10	Andrology	35	Chemistry, Applied
11	Anesthesiology	36	Chemistry, Inorganic & Nuclear
12	Anthropology	37	Chemistry, Medicinal
13	Applied Linguistics	38	Chemistry, Multidisciplinary
14	Archeology	39	Chemistry, Organic
15	Architecture	40	Chemistry, Physical
16	Area Studies	41	Classics
17	Art	42	Clinical Neurology
18	Arts & Humanities, General	43	Communication
19	Asian Studies	44	Computer Science, Artificial Intelligence
20	Astronomy & Astrophysics	45	Computer Science, Cybernetics
21	Automation & Control Systems	46	Computer Science, Hardware & Architecture
22	Behavioral Sciences	47	Computer Science, Information Systems
23	Biochemical Research Methods	48	Computer Science, Interdisciplinary Applications
24	Biochemistry & Molecular Biology	49	Computer Science, Software, Graphics, Programming
25	Biodiversity Conservation	50	Computer Science, Theory & Methods

51 Construction & Building Technology
52 Criminology & Penology
53 Critical Care Medicine
54 Crystallography
55 Dance
56 Demography
57 Dentistry, Oral Surgery & Medicine
58 Dermatology & Venereal Diseases
59 Developmental Biology
60 Ecology
61 Economics
62 Education & Educational Research
63 Education, Scientific Disciplines
64 Education, Special
65 Electrochemistry
66 Emergency Medicine & Critical Care
67 Endocrinology & Metabolism
68 Energy & Fuels
69 Engineering
70 Engineering, Aerospace
71 Engineering, Biomedical
72 Engineering, Chemical
73 Engineering, Civil
74 Engineering, Electrical & Electronic
75 Engineering, Environmental
76 Engineering, Geological
77 Engineering, Industrial
78 Engineering, Manufacturing
79 Engineering, Marine
80 Engineering, Mechanical
81 Engineering, Ocean
82 Engineering, Petroleum
83 Entomology
84 Environmental Sciences
85 Environmental Studies

86 Ergonomics
87 Ethics
88 Ethnic Studies
89 Evolutionary Biology
90 Family Studies
91 Film, Radio, Television
92 Fisheries
93 Folklore
94 Food Science & Technology
95 Forestry
96 Gastroenterology & Hepatology
97 Genetics & Heredity
98 Geochemistry & Geophysics
99 Geography
100 Geography, Physical
101 Geology
102 Geosciences, Interdisciplinary
103 Geriatrics & Gerontology
104 Gerontology
105 Health Care Sciences & Services
106 Health Policy & Services
107 Hematology
108 History
109 History & Philosophy Of Science
110 History Of Social Sciences
111 Horticulture
112 Imaging Science & Photographic Technology
113 Immunology
114 Industrial Relations & Labor
115 Infectious Diseases
116 Information Science & Library Science
117 Instruments & Instrumentation
118 Integrative & Complementary Medicine
119 International Relations
120 Language & Linguistics

121	Law	151	Medicine, Research & Experimental
122	Limnology	152	Metallurgy & Metallurgical Engineering
123	Literary Reviews	153	Meteorology & Atmospheric Sciences
124	Literary Theory & Criticism	154	Microbiology
125	Literature	155	Microscopy
126	Literature, African, Australian, Canadian	156	Mineralogy
127	Literature, American	157	Mining & Mineral Processing
128	Literature, British Isles	158	Music
129	Literature, German, Netherlandic, Scandinavian	159	Mycology
130	Literature, Romance	160	Neuroimaging
131	Literature, Slavic	161	Neurosciences
132	Management	162	Nuclear Science & Technology
133	Marine & Freshwater Biology	163	Nursing
134	Materials Science, Biomaterials	164	Nutrition & Dietetics
135	Materials Science, Ceramics	165	Obstetrics & Gynecology
136	Materials Science, Characterization & Testing	166	Oceanography
137	Materials Science, Coatings & Films	167	Oncology
138	Materials Science, Composites	168	Operations Research & Management Science
139	Materials Science, Multidisciplinary	169	Ophthalmology
140	Materials Science, Paper & Wood	170	Optics
141	Materials Science, Textiles	171	Ornithology
142	Mathematics	172	Orthopedics
143	Mathematics, Applied	173	Otorhinolaryngology
144	Mathematics, Miscellaneous	174	Paleontology
145	Mechanics	175	Parasitology
146	Medical Ethics	176	Pathology
147	Medical Informatics	177	Pediatrics
148	Medical Laboratory Technology	178	Peripheral Vascular Disease
149	Medicine, General & Internal	179	Pharmacology & Pharmacy
150	Medicine, Legal	180	Philosophy

181 Physics, Applied	212 Remote Sensing
182 Physics, Atomic, Molecular & Chemical	213 Reproductive Systems
183 Physics, Condensed Matter	214 Respiratory System
184 Physics, Fluids & Plasmas	215 Rheumatology
185 Physics, Mathematical	216 Robotics
186 Physics, Multidisciplinary	217 Social Issues
187 Physics, Nuclear	218 Social Sciences, Biomedical
188 Physics, Particles & Fields	219 Social Sciences, Interdisciplinary
189 Physiology	220 Social Sciences, Mathematical Methods
190 Planning & Development	221 Social Work
191 Plant Sciences	222 Sociology
192 Poetry	223 Spectroscopy
193 Political Science	224 Sport Sciences
194 Polymer Science	225 Statistics & Probability
195 Psychiatry	226 Substance Abuse
196 Psychology	227 Surgery
197 Psychology, Applied	228 Telecommunications
198 Psychology, Biological	229 Theater
199 Psychology, Clinical	230 Thermodynamics
200 Psychology, Developmental	231 Toxicology
201 Psychology, Educational	232 Transplantation
202 Psychology, Experimental	233 Transportation
203 Psychology, Mathematical	234 Tropical Medicine
204 Psychology, Multidisciplinary	235 Urban Studies
205 Psychology, Psychoanalysis	236 Urology & Nephrology
206 Psychology, Social	237 Veterinary Sciences
207 Public Administration	238 Virology
208 Public, Environmental & Occupational Health	239 Water Resources
209 Radiology, Nuclear Medicine & Medical Imaging	240 Women's Studies
210 Rehabilitation	241 Zoology
211 Religion	

Annex II

Categories With Links Coinciding in the Maps of Spain 2002 and the JCR, Based on Their Journals

Agricultural Economics & Policy	Economics
Agriculture	Agriculture, Soil Science
Agriculture	Plant Sciences
Anesthesiology	Clinical Neurology
Anthropology	Archeology
Archeology	Architecture
Archeology	Art
Behavioral Sciences	Psychology, Biological
Biochemical Research Methods	Chemistry, Analytical
Biochemistry & Molecular Biology	Biochemical Research Methods
Biochemistry & Molecular Biology	Biophysics
Biochemistry & Molecular Biology	Endocrinology & Metabolism
Biochemistry & Molecular Biology	Genetics & Heredity
Biochemistry & Molecular Biology	Medical Laboratory Technology
Biochemistry & Molecular Biology	Pharmacology & Pharmacy
Biochemistry & Molecular Biology	Physiology
Biochemistry & Molecular Biology	Plant Sciences
Business	Management
Cardiac & Cardiovascular Systems	Respiratory System
Chemistry, Applied	Materials Science, Textiles
Clinical Neurology	Psychiatry
Computer Science, Information Systems	Computer Science, Software, Graphics, Programming
Computer Science, Information Systems	Information Science & Library Science
Computer Science, Theory & Methods	Computer Science, Software, Graphics, Programming
Crystallography	Chemistry, Inorganic & Nuclear
Economics	Business, Finance
Economics	Industrial Relations & Labor
Economics	International Relations
Economics	Social Sciences, Mathematical Methods
Education & Educational Research	Psychology, Educational

Energy & Fuels	Engineering, Chemical
Energy & Fuels	Engineering, Petroleum
Engineering, Biomedical	Materials Science, Biomaterials
Engineering, Civil	Engineering, Ocean
Engineering, Electrical & Electronic	Automation & Control Systems
Engineering, Electrical & Electronic	Computer Science, Artificial Intelligence
Engineering, Electrical & Electronic	Telecommunications
Entomology	Biochemistry & Molecular Biology
Environmental Sciences	Engineering, Environmental
Environmental Sciences	Water Resources
Environmental Studies	Urban Studies
Ergonomics	Transportation
Food Science & Technology	Biotechnology & Applied Microbiology
Food Science & Technology	Chemistry, Applied
Forestry	Materials Science, Paper & Wood
Genetics & Heredity	Biology, Miscellaneous
Geochemistry & Geophysics	Mineralogy
Geosciences, Interdisciplinary	Engineering, Geological
Geosciences, Interdisciplinary	Geography
Geosciences, Interdisciplinary	Meteorology & Atmospheric Sciences
Health Policy & Services	Health Care Sciences & Services
Hematology	Peripheral Vascular Disease
History	History Of Social Sciences
History	Literature
History	Religion
Immunology	Allergy
Immunology	Infectious Diseases
Language & Linguistics	Literature
Language & Linguistics	Literature, Romance
Marine & Freshwater Biology	Ecology
Marine & Freshwater Biology	Fisheries
Materials Science, Multidisciplinary	Chemistry, Physical
Materials Science, Multidisciplinary	Physics, Condensed Matter
Mathematics	Mathematics, Applied
Mathematics, Applied	Engineering
Mathematics, Applied	Mathematics, Miscellaneous
Mathematics, Applied	Operations Research & Management Science
Mathematics, Applied	Physics, Mathematical
Mathematics, Miscellaneous	Psychology, Mathematical
Mathematics, Miscellaneous	Social Sciences, Mathematical Methods
Mechanics	Engineering, Mechanical
Mechanics	Physics, Fluids & Plasmas
Neurosciences	Behavioral Sciences
Neurosciences	Clinical Neurology
Obstetrics & Gynecology	Reproductive Biology

Operations Research & Management Science	Engineering, Industrial
Pharmacology & Pharmacy	Chemistry, Medicinal
Pharmacology & Pharmacy	Toxicology
Physics, Applied	Engineering, Electrical & Electronic
Physics, Atomic, Molecular & Chemical	Chemistry, Physical
Physics, Multidisciplinary	Physics, Mathematical
Psychiatry	Psychology
Psychology	Psychology, Applied
Psychology	Psychology, Experimental
Public, Environmental & Occupational Health	Social Sciences, Biomedical
Radiology, Nuclear Medicine & Medical Imaging	Acoustics
Rehabilitation	Education, Special
Rehabilitation	Sport Sciences
Remote Sensing	Imaging Science & Photographic Technology
Statistics & Probability	Mathematics, Miscellaneous
Surgery	Dentistry, Oral Surgery & Medicine
Surgery	Orthopedics
Surgery	Otorhinolaryngology
Tropical Medicine	Public, Environmental & Occupational Health
Water Resources	Engineering, Civil

Categories With Links Not Coinciding in the Maps of Spain 2002 and the JCR, Based on Their Journals

Agriculture, Dairy & Animal Science	Food Science & Technology
Anatomy & Morphology	Neurosciences
Anesthesiology	Poetry
Area Studies	History
Astronomy & Astrophysics	Meteorology & Atmospheric Sciences
Biochemistry & Molecular Biology	Biology
Biochemistry & Molecular Biology	Biotechnology & Applied Microbiology
Biochemistry & Molecular Biology	Cell Biology
Biochemistry & Molecular Biology	Developmental Biology
Biochemistry & Molecular Biology	Geriatrics & Gerontology
Biochemistry & Molecular Biology	Immunology

Biochemistry & Molecular Biology	Medicine, Research & Experimental
Biochemistry & Molecular Biology	Microbiology
Biochemistry & Molecular Biology	Microscopy
Biochemistry & Molecular Biology	Mycology
Biochemistry & Molecular Biology	Neurosciences
Biochemistry & Molecular Biology	Oncology
Biochemistry & Molecular Biology	Parasitology
Biochemistry & Molecular Biology	Reproductive Biology
Biology, Miscellaneous	Anthropology
Cardiac & Cardiovascular Systems	Peripheral Vascular Disease
Cell Biology	Pathology
Chemistry, Analytical	Electrochemistry
Chemistry, Analytical	Philosophy
Chemistry, Analytical	Spectroscopy
Chemistry, Multidisciplinary	Biochemistry & Molecular Biology
Chemistry, Multidisciplinary	Chemistry, Inorganic & Nuclear
Chemistry, Multidisciplinary	Chemistry, Organic
Chemistry, Multidisciplinary	Chemistry, Physical
Computer Science, Artificial Intelligence	Computer Science, Cybernetics
Computer Science, Interdisciplinary Applications	Physics, Mathematical
Computer Science, Software, Graphics, Programming	Computer Science, Hardware & Architecture
Computer Science, Theory & Methods	Computer Science, Artificial Intelligence
Construction & Building Technology	Materials Science, Multidisciplinary
Ecology	Biology, Miscellaneous
Ecology	Ornithology
Economics	Management
Economics	Planning & Development
Economics	Political Science
Education, Scientific Disciplines	Physics, Multidisciplinary
Engineering, Aerospace	Astronomy & Astrophysics
Engineering, Chemical	Chemistry, Physical
Engineering, Chemical	Thermodynamics
Engineering, Manufacturing	Engineering, Industrial
Environmental Sciences	Chemistry, Analytical
Environmental Studies	Economics
Gastroenterology & Hepatology	Medicine, General & Internal
Genetics & Heredity	History
Geochemistry & Geophysics	Remote Sensing
Geosciences, Interdisciplinary	Geochemistry & Geophysics
Geosciences, Interdisciplinary	Geology

Geosciences, Interdisciplinary	Oceanography
Geosciences, Interdisciplinary	Paleontology
History & Philosophy Of Science	Philosophy
Immunology	Transplantation
Instruments & Instrumentation	Nuclear Science & Technology
Instruments & Instrumentation	Spectroscopy
Law	Economics
Literary Reviews	Literature, American
Literature	Arts & Humanities, General
Marine & Freshwater Biology	Oceanography
Materials Science, Multidisciplinary	Materials Science, Ceramics
Materials Science, Multidisciplinary	Materials Science, Composites
Materials Science, Multidisciplinary	Metallurgy &Amp; Metallurgical Engineering
Materials Science, Multidisciplinary	Mining & Mineral Processing
Medical Informatics	Health Care Sciences & Services
Medicine, General & Internal	Biochemistry & Molecular Biology
Medicine, General & Internal	Cardiac & Cardiovascular Systems
Medicine, General & Internal	Dermatology & Venereal Diseases
Medicine, General & Internal	Emergency Medicine & Critical Care
Medicine, General & Internal	Health Care Sciences & Services
Medicine, General & Internal	Pediatrics
Medicine, General & Internal	Public, Environmental & Occupational Health
Medicine, General & Internal	Rheumatology
Medicine, General & Internal	Surgery
Medicine, General & Internal	Urology & Nephrology
Music	Art
Neurosciences	Ophthalmology
Nursing	Health Care Sciences & Services
Nutrition & Dietetics	Biochemistry & Molecular Biology
Oceanography	Engineering, Marine
Oceanography	Limnology
Optics	Physics, Atomic, Molecular & Chemical
Pathology	Medicine, Legal
Physics, Applied	Physics, Condensed Matter
Physics, Condensed Matter	Materials Science, Coatings & Films
Physics, Fluids & Plasmas	Physics, Mathematical
Physics, Multidisciplinary	Physics, Condensed Matter
Physics, Multidisciplinary	Physics, Nuclear
Physics, Multidisciplinary	Physics, Particles & Fields
Physiology	Sport Sciences

Plant Sciences	Forestry
Plant Sciences	Horticulture
Political Science	Theater
Polymer Science	Chemistry, Physical
Polymer Science	Materials Science, Characterization & Testing
Psychiatry	Psychology, Developmental
Psychiatry	Substance Abuse
Psychology	Ergonomics
Psychology	Psychology, Clinical
Psychology	Psychology, Educational
Psychology	Psychology, Psychoanalysis
Psychology	Psychology, Social
Psychology	Social Sciences, Interdisciplinary
Psychology	Social Work
Psychology	Sociology
Psychology, Social	Family Studies
Public Administration	Political Science
Public, Environmental & Occupational Health	Women's Studies
Radiology, Nuclear Medicine & Medical Imaging	Medicine, General & Internal
Reproductive Biology	Andrology
Social Sciences, Biomedical	Demography
Social Sciences, Biomedical	Social Issues
Social Sciences, Interdisciplinary	Communication
Surgery	Engineering, Biomedical
Theater	Literary Reviews
Veterinary Sciences	Immunology
Virology	Immunology
Zoology	Neurosciences

Categories With Links Coinciding in the Maps of Spain 2002 and the JCR, Based on Their Document Inter-Citation

Anatomy & Morphology	Neurosciences
Anesthesiology	Poetry
Archeology	Art
Area Studies	History
Astronomy & Astrophysics	Meteorology & Atmospheric Sciences
Behavioral Sciences	Psychology, Biological
Biochemistry & Molecular Biology	Cell Biology

Biochemistry & Molecular Biology	Developmental Biology
Biochemistry & Molecular Biology	Geriatrics & Gerontology
Biochemistry & Molecular Biology	Immunology
Biochemistry & Molecular Biology	Medical Laboratory Technology
Biochemistry & Molecular Biology	Medicine, Research & Experimental
Biochemistry & Molecular Biology	Microbiology
Biochemistry & Molecular Biology	Microscopy
Biochemistry & Molecular Biology	Mycology
Biochemistry & Molecular Biology	Neurosciences
Biochemistry & Molecular Biology	Oncology
Biochemistry & Molecular Biology	Parasitology
Biochemistry & Molecular Biology	Physiology
Biochemistry & Molecular Biology	Reproductive Biology
Chemistry, Analytical	Electrochemistry
Chemistry, Analytical	Philosophy
Chemistry, Analytical	Spectroscopy
Chemistry, Multidisciplinary	Biochemistry & Molecular Biology
Chemistry, Multidisciplinary	Chemistry, Inorganic & Nuclear
Chemistry, Multidisciplinary	Chemistry, Organic
Chemistry, Multidisciplinary	Chemistry, Physical
Computer Science, Information Systems	Computer Science, Software, Graphics, Programming
Computer Science, Interdisciplinary Applications	Physics, Mathematical
Computer Science, Theory & Methods	Computer Science, Artificial Intelligence
Construction & Building Technology	Materials Science, Multidisciplinary
Ecology	Biology, Miscellaneous
Ecology	Ornithology
Economics	International Relations
Economics	Management
Education, Scientific Disciplines	Physics, Multidisciplinary
Engineering, Chemical	Thermodynamics
Environmental Sciences	Chemistry, Analytical
Environmental Studies	Economics
Gastroenterology & Hepatology	Medicine, General & Internal
Genetics & Heredity	Biology, Miscellaneous
Genetics & Heredity	History
Geosciences, Interdisciplinary	Geochemistry & Geophysics
Geosciences, Interdisciplinary	Geography
Geosciences, Interdisciplinary	Geology
History & Philosophy of Science	Philosophy
Law	Economics
Literary Reviews	Literature, American
Literature	Arts & Humanities, General
Materials Science, Multidisciplinary	Materials Science, Composites

Materials Science, Multidisciplinary	Metallurgy & Metallurgical Engineering
Mathematics, Applied	Engineering
Mathematics, Applied	Mathematics, Miscellaneous
Mathematics, Applied	Operations Research & Management Science
Medical Informatics	Health Care Sciences & Services
Medicine, General & Internal	Biochemistry & Molecular Biology
Medicine, General & Internal	Cardiac & Cardiovascular Systems
Medicine, General & Internal	Dermatology & Venereal Diseases
Medicine, General & Internal	Emergency Medicine & Critical Care
Medicine, General & Internal	Health Care Sciences & Services
Medicine, General & Internal	Pediatrics
Medicine, General & Internal	Rheumatology
Medicine, General & Internal	Surgery
Medicine, General & Internal	Urology & Nephrology
Music	Art
Neurosciences	Ophthalmology
Nutrition & Dietetics	Biochemistry & Molecular Biology
Oceanography	Engineering, Marine
Oceanography	Limnology
Physics, Applied	Physics, Condensed Matter
Physics, Atomic, Molecular & Chemical	Chemistry, Physical
Physics, Condensed Matter	Materials Science, Coatings & Films
Physics, Multidisciplinary	Physics, Condensed Matter
Physics, Multidisciplinary	Physics, Nuclear
Physics, Multidisciplinary	Physics, Particles & Fields
Plant Sciences	Forestry
Plant Sciences	Horticulture
Political Science	Theater
Polymer Science	Chemistry, Physical
Polymer Science	Materials Science, Characterization & Testing
Psychiatry	Psychology
Psychiatry	Psychology, Developmental
Psychology	Ergonomics
Psychology	Psychology, Clinical
Psychology	Psychology, Educational
Psychology	Psychology, Psychoanalysis
Psychology	Psychology, Social
Psychology	Social Sciences, Interdisciplinary
Psychology	Social Work
Psychology	Sociology
Psychology, Social	Family Studies
Public Administration	Political Science

Radiology, Nuclear Medicine & Medical Imaging	Medicine, General & Internal
Reproductive Biology	Andrology
Social Sciences, Biomedical	Demography
Social Sciences, Biomedical	Social Issues
Social Sciences, Interdisciplinary	Communication
Statistics & Probability	Mathematics, Miscellaneous
Surgery	Engineering, Biomedical
Theater	Literary Reviews
Virology	Immunology
Zoology	Neurosciences

Categories With Links Not Coinciding in the Maps of Spain 2002 and the JCR, Based on Their Document Inter-Citation

Agricultural Economics & Policy	Economics
Agriculture	Agriculture, Soil Science
Agriculture	Plant Sciences
Agriculture, Dairy & Animal Science	Food Science & Technology
Anesthesiology	Clinical Neurology
Anthropology	Archeology
Archeology	Architecture
Biochemical Research Methods	Chemistry, Analytical
Biochemistry & Molecular Biology	Biochemical Research Methods
Biochemistry & Molecular Biology	Biology
Biochemistry & Molecular Biology	Biophysics
Biochemistry & Molecular Biology	Biotechnology & Applied Microbiology
Biochemistry & Molecular Biology	Endocrinology & Metabolism
Biochemistry & Molecular Biology	Genetics & Heredity
Biochemistry & Molecular Biology	Pharmacology & Pharmacy
Biochemistry & Molecular Biology	Plant Sciences
Biology, Miscellaneous	Anthropology
Business	Management
Cardiac & Cardiovascular Systems	Peripheral Vascular Disease
Cardiac & Cardiovascular Systems	Respiratory System
Cell Biology	Pathology
Chemistry, Applied	Materials Science, Textiles
Clinical Neurology	Psychiatry
Computer Science, Artificial Intelligence	Computer Science, Cybernetics
Computer Science, Information Systems	Information Science & Library Science

Computer Science, Software, Graphics, Programming	Computer Science, Hardware & Architecture
Computer Science, Theory & Methods	Computer Science, Software, Graphics, Programming
Crystallography	Chemistry, Inorganic & Nuclear
Economics	Business, Finance
Economics	Industrial Relations & Labor
Economics	Planning & Development
Economics	Political Science
Economics	Social Sciences, Mathematical Methods
Education & Educational Research	Psychology, Educational
Energy & Fuels	Engineering, Chemical
Energy & Fuels	Engineering, Petroleum
Engineering, Aerospace	Astronomy & Astrophysics
Engineering, Biomedical	Materials Science, Biomaterials
Engineering, Chemical	Chemistry, Physical
Engineering, Civil	Engineering, Ocean
Engineering, Electrical & Electronic	Automation & Control Systems
Engineering, Electrical & Electronic	Computer Science, Artificial Intelligence
Engineering, Electrical & Electronic	Telecommunications
Engineering, Manufacturing	Engineering, Industrial
Entomology	Biochemistry & Molecular Biology
Environmental Sciences	Engineering, Environmental
Environmental Sciences	Water Resources
Environmental Studies	Urban Studies
Ergonomics	Transportation
Food Science & Technology	Biotechnology & Applied Microbiology
Food Science & Technology	Chemistry, Applied
Forestry	Materials Science, Paper & Wood
Geochemistry & Geophysics	Mineralogy
Geochemistry & Geophysics	Remote Sensing
Geosciences, Interdisciplinary	Engineering, Geological
Geosciences, Interdisciplinary	Meteorology & Atmospheric Sciences
Geosciences, Interdisciplinary	Oceanography
Geosciences, Interdisciplinary	Paleontology
Health Policy & Services	Health Care Sciences & Services
Hematology	Peripheral Vascular Disease
History	History Of Social Sciences
History	Literature
History	Religion
Immunology	Allergy
Immunology	Infectious Diseases
Immunology	Transplantation
Instruments & Instrumentation	Nuclear Science & Technology
Instruments & Instrumentation	Spectroscopy
Language & Linguistics	Literature

Language & Linguistics	Literature, Romance
Marine & Freshwater Biology	Ecology
Marine & Freshwater Biology	Fisheries
Marine & Freshwater Biology	Oceanography
Materials Science, Multidisciplinary	Chemistry, Physical
Materials Science, Multidisciplinary	Materials Science, Ceramics
Materials Science, Multidisciplinary	Mining & Mineral Processing
Materials Science, Multidisciplinary	Physics, Condensed Matter
Mathematics	Mathematics, Applied
Mathematics, Applied	Physics, Mathematical
Mathematics, Miscellaneous	Psychology, Mathematical
Mathematics, Miscellaneous	Social Sciences, Mathematical Methods
Mechanics	Engineering, Mechanical
Mechanics	Physics, Fluids & Plasmas
Medicine, General & Internal	Public, Environmental & Occupational Health
Neurosciences	Behavioral Sciences
Neurosciences	Clinical Neurology
Nursing	Health Care Sciences & Services
Obstetrics & Gynecology	Reproductive Biology
Operations Research & Management Science	Engineering, Industrial
Optics	Physics, Atomic, Molecular & Chemical
Pathology	Medicine, Legal
Pharmacology & Pharmacy	Chemistry, Medicinal
Pharmacology & Pharmacy	Toxicology
Physics, Applied	Engineering, Electrical & Electronic
Physics, Fluids & Plasmas	Physics, Mathematical
Physics, Multidisciplinary	Physics, Mathematical
Physiology	Sport Sciences
Psychiatry	Substance Abuse
Psychology	Psychology, Applied
Psychology	Psychology, Experimental
Public, Environmental & Occupational Health	Social Sciences, Biomedical
Public, Environmental & Occupational Health	Women's Studies
Radiology, Nuclear Medicine & Medical Imaging	Acoustics
Rehabilitation	Education, Special
Rehabilitation	Sport Sciences
Remote Sensing	Imaging Science & Photographic Technology
Surgery	Dentistry, Oral Surgery & Medicine
Surgery	Orthopedics

Surgery	Otorhinolaryngology
Tropical Medicine	Public, Environmental & Occupational Health
Veterinary Sciences	Immunology
Water Resources	Engineering, Civil

Annex III

Factors Extracted from EU, 2002

Factor 1	% variance	Factor 2	% variance
Biomedicine	19.6	Materials Science & Applied Physics	10.5
Medicine, Research & Experimental	0.947	*Materials Science, Ceramics*	0.964
Endocrinology & Metabolism	0.943	*Metallurgy & Metallurgical Engineering*	0.927
Oncology	0.934	*Materials Science, Coatings & Films*	0.898
Pathology	0.926	*Polymer Science*	0.818
Medical Laboratory Technology	0.922	*Physics, Applied*	0.815
Urology & Nephrology	0.907	*Physics, Condensed Matter*	0.753
Physiology	0.898	*Materials Science, Multidisciplinary*	0.746
Dermatology & Venereal Diseases	0.894	*Electrochemistry*	0.746
Immunology	0.887	*Chemistry, Physical*	0.740
Gastroenterology & Hepatology	0.882	*Materials Science, Characterization & Testing*	0.735
Pharmacology & Pharmacy	0.876	**Mining & Mineral Processing**	0.670
Biology	0.867	**Physics, Atomic, Molecular & Chemical**	0.664
Biophysics	0.864	**Engineering, Electrical & Electronic**	0.634
Developmental Biology	0.859	**Crystallography**	0.626
Pediatrics	0.854	**Materials Science, Composites**	0.616
Genetics & Heredity	0.851	**Mechanics**	0.614
Nutrition & Dietetics	0.851	**Physics, Multidisciplinary**	0.604

Factor 1	% variance	Factor 2	% variance
Rheumatology	0.845	**Optics**	0.592
Geriatrics & Gerontology	0.837	**Engineering, Chemical**	0.555
Cell Biology	0.834	**Energy & Fuels**	0.537
Anatomy & Morphology	0.828	**Chemistry, Multidisciplinary**	0.531
Ophthalmology	0.820	**Construction & Building Technology**	0.502
Virology	0.814		
Medicine, General & Internal	0.813		
Hematology	0.805		
Biotechnology & Applied Microbiology	0.797		
Toxicology	0.794		
Andrology	0.778		
Veterinary Sciences	0.776		
Microbiology	0.760		
Neurosciences	0.758		
Biochemical Research Methods	0.750		
Obstetrics & Gynecology	0.737		
Reproductive Biology	0.736		
Respiratory System	0.725		
Biochemistry & Molecular Biology	0.722		
Microscopy	0.714		
Dentistry, Oral Surgery & Medicine	0.711		
Surgery	0.705		
Medicine, Legal	0.697		
Radiology, Nuclear Medicine & Medical Imaging	0.673		
Parasitology	0.669		
Public, Environmental & Occupational Health	0.668		
Plant Sciences	0.666		
Peripheral Vascular Disease	0.663		
Transplantation	0.647		
Cardiac & Cardiovascular Systems	0.644		
Chemistry, Analytical	0.643		
Allergy	0.636		
Mycology	0.616		
Infectious Diseases	0.607		
Otorhinolaryngology	0.606		
Zoology	0.595		

Factor 1	% variance	Factor 2	% variance
Chemistry, Medicinal	0.588		
Clinical Neurology	0.577		
Sport Sciences	0.571		
Agriculture, Dairy & Animal Science	0.569		
Biology, Miscellaneous	0.552		
Anesthesiology	0.533		
Engineering, Biomedical	0.509		

Factor 3	% variance	Factor 4	% variance
Management, Law & Economy	6.9	Earth & Space Sciences	6.1
Planning & Development	0.876	*Geology*	0.895
Industrial Relations & Labor	0.870	*Paleontology*	0.892
Political Science	0.841	*Oceanography*	0.888
International Relations	0.824	*Meteorology & Atmospheric Sciences*	0.883
Public Administration	0.820	*Astronomy & Astrophysics*	0.870
Area Studies	0.820	*Geochemistry & Geophysics*	0.853
Sociology	0.810	*Engineering, Ocean*	0.807
Agricultural Economics & Policy	0.786	*Engineering, Aerospace*	0.799
History Of Social Sciences	0.726	*Geosciences, Interdisciplinary*	0.750
Urban Studies	0.682	*Engineering, Geological*	0.746
Business, Finance	0.680	*Remote Sensing*	0.742
Law	0.669	*Mineralogy*	0.721
Social Issues	0.641	*Geography*	0.703
Environmental Studies	0.610	**Limnology**	0.625
Ethnic Studies	0.588	**Engineering, Petroleum**	0.585
Business	0.571	**Environmental Sciences**	0.573
Management	0.559	**Water Resources**	0.545
Arts & Humanities, General	0.554	**Engineering, Civil**	0.533
Social Sciences, Interdisciplinary	0.553		
History	0.526		
Communication	0.525		
Demography	0.521		

Factor 5	% variance	Factor 6	% variance
Psychology	4.8	Computer Science & Telecommunications	3.6
Psychology, Social	0.910	*Computer Science, Hardware & Architecture*	0.911
Psychology, Developmental	0.871	*Computer Science, Artificial Intelligence*	0.883
Psychology, Clinical	0.863	*Computer Science, Theory & Methods*	0.839
Psychology, Psychoanalysis	0.852	*Computer Science, Software, Graphics, Programming*	0.839
Psychology, Experimental	0.757	*Computer Science, Information Systems*	0.833
Psychology, Educational	0.743	*Automation & Control Systems*	0.743
Criminology & Penology	0.739	*Telecommunications*	0.724
Psychology, Applied	0.732	*Computer Science, Cybernetics*	0.720
Family Studies	0.718	**Operations Research & Management Science**	0.588
Psychology	0.692	**Information Science & Library Science**	0.513
Substance Abuse	0.676		
Education & Educational Research	0.650		
Social Work	0.643		
Psychology, Biological	0.637		
Psychiatry	0.582		
Behavioral Sciences	0.580		
Education, Special	0.577		
Women's Studies	0.544		
Social Sciences, Interdisciplinary	0.522		
Ergonomics	0.510		
Psychology, Mathematical	0.507		
Communication	0.503		

Factor 7	% variance	Factor 8	% variance
Animal Biology, Ecology	3.2	Humanities	2.4
Ornithology	0.802	*Literature, Romance*	0.877
Entomology	0.693	*Literature*	0.825
Zoology	0.627	**Language & Linguistics**	0.659
Anthropology	0.620	**Arts & Humanities, General**	0.649
Ecology	0.616	**Literary Reviews**	0.642
Fisheries	0.555	**Religion**	0.632
Marine & Freshwater Biology	0.524	**Literature, Slavic**	0.605
Biology, Miscellaneous	0.522		

Factor 9	% variance	Factor 10	% variance
Physics, Particles & Fields	2.1	Health Policy & Services	1.9
Physics, Nuclear	0.849	*Social Sciences, Biomedical*	0.823
Physics, Particles & Fields	0.829	*Nursing*	0.799
Physics, Mathematical	0.799	*Health Policy & Services*	0.799
Nuclear Science & Technology	0.712	**Health Care Sciences & Services**	0.686
Instruments & Instrumentation	0.696	**Tropical Medicine**	0.652
Computer Science, Interdisciplinary Applications	0.681	**Medical Informatics**	0.581
Spectroscopy	0.673	**Women's Studies**	0.568
Physics, Fluids & Plasmas	0.666	**Social Work**	0.552
Education, Scientific Disciplines	0.592		
Optics	0.589		
Physics, Multidisciplinary	0.501		

Factor 11	% variance	Factor 12	% variance
Engineering Mechanical	1.6	Orthopedics	1.4
Engineering, Mechanical	0.739	*Emergency Medicine & Critical Care*	0.740
Engineering	0.689	*Orthopedics*	0.726
Thermodynamics	0.620	**Otorhinolaryngology**	0.595
Materials Science, Composites	0.534	**Anesthesiology**	0.506
Acoustics	0.504		

Factor 13	% variance	Factor 14	% variance
Applied Mathematics	1.2	Chemistry	1.1
Social Sciences, Mathematical Methods	0.754	*Chemistry, Organic*	0.738
Statistics & Probability	0.697	**Chemistry, Applied**	0.694
Psychology, Mathematical	0.662	**Chemistry, Inorganic & Nuclear**	0.656
Economics	0.636	**Chemistry, Medicinal**	0.609
Mathematics, Miscellaneous	0.632	**Engineering, Chemical**	0.596
Business, Finance	0.623	**Materials Science, Textiles**	0.555
		Crystallography	0.548
		Chemistry, Multidisciplinary	0.520

Factor 15	% variance
Agriculture & Soil Sciences	1
Horticulture	0.814
Agriculture	0.759
Forestry	0.719
Agriculture, Soil Science	0.650

Factors Extracted from US, 2002

Factor 1	% variance	Factor 2	% variance
Biomedicine	20.5	Psychology	11.7
Endocrinology & Metabolism	0.955	*Psychology, Social*	0.908
Medicine, Research & Experimental	0.949	*Psychology, Developmental*	0.868
Pathology	0.942	*Social Work*	0.832
Oncology	0.940	*Family Studies*	0.831
Urology & Nephrology	0.937	*Psychology, Clinical*	0.826
Physiology	0.911	*Education & Educational Research*	0.821
Gastroenterology & Hepatology	0.909	*Psychology, Educational*	0.819
Medical Laboratory Technology	0.902	*Psychology, Applied*	0.804
Pharmacology & Pharmacy	0.902	*Psychology, Psychoanalysis*	0.802
Ophthalmology	0.899	*Women's Studies*	0.802
Biophysics	0.896	*Criminology & Penology*	0.801
Immunology	0.895	*Substance Abuse*	0.737
Dermatology & Venereal Diseases	0.887	*Psychology, Experimental*	0.731
Biology	0.883	*Psychology*	0.701
Genetics & Heredity	0.875	**Social Sciences, Interdisciplinary**	0.694
Developmental Biology	0.875	**Psychiatry**	0.621
Rheumatology	0.867	**Social Issues**	0.614
Anatomy & Morphology	0.861	**Social Sciences, Biomedical**	0.582
Geriatrics & Gerontology	0.852	**Sociology**	0.576
Biotechnology & Applied Microbiology	0.848	**Communication**	0.568
Nutrition & Dietetics	0.845	**Psychology, Mathematical**	0.563
Virology	0.841	**Ergonomics**	0.549
Cell Biology	0.838	**Education, Special**	0.527
Hematology	0.835	**Psychology, Biological**	0.517
Toxicology	0.826	**Religion**	0.503
Neurosciences	0.816	**Rehabilitation**	0.501
Andrology	0.808		
Medicine, General & Internal	0.806		
Respiratory System	0.803		
Biochemical Research Methods	0.802		

Factor 1	% variance	Factor 2	% variance
Dentistry, Oral Surgery & Medicine	0.799		
Reproductive Biology	0.797		
Pediatrics	0.789		
Microbiology	0.781		
Microscopy	0.775		
Veterinary Sciences	0.763		
Obstetrics & Gynecology	0.758		
Surgery	0.747		
Biochemistry & Molecular Biology	0.747		
Radiology, Nuclear Medicine & Medical Imaging	0.724		
Peripheral Vascular Disease	0.693		
Chemistry, Analytical	0.692		
Parasitology	0.692		
Plant Sciences	0.683		
Cardiac & Cardiovascular Systems	0.675		
Chemistry, Medicinal	0.671		
Mycology	0.657		
Transplantation	0.652		
Allergy	0.641		
Sport Sciences	0.629		
Clinical Neurology	0.627		
Otorhinolaryngology	0.624		
Engineering, Biomedical	0.617		
Zoology	0.605		
Infectious Diseases	0.591		
Food Science & Technology	0.579		
Public, Environmental & Occupational Health	0.567		
Agriculture, Dairy & Animal Science	0.566		
Anesthesiology	0.561		
Biology, Miscellaneous	0.546		
Entomology	0.529		
Emergency Medicine & Critical Care	0.509		
Crystallography	0.508		

Factor 3	% variance	Factor 4	% variance
Materials Science & Applied Physics	7.3	Earth & Space Sciences	6.1
Materials Science, Ceramics	0.952	*Geology*	0.927
Materials Science, Coatings & Films	0.890	*Paleontology*	0.897
Metallurgy & Metallurgical Engineering	0.876	*Astronomy & Astrophysics*	0.897
Physics, Condensed Matter	0.829	*Oceanography*	0.892
Physics, Applied	0.825	*Meteorology & Atmospheric Sciences*	0.889
Polymer Science	0.822	*Engineering, Ocean*	0.853
Materials Science, Multidisciplinary	0.793	*Geochemistry & Geophysics*	0.850
Electrochemistry	0.792	*Geosciences, Interdisciplinary*	0.760
Chemistry, Physical	0.778	*Remote Sensing*	0.756
Physics, Atomic, Molecular & Chemical	0.743	*Mineralogy*	0.746
Mining & Mineral Processing	0.732	*Engineering, Geological*	0.728
Physics, Multidisciplinary	0.722	*Engineering, Aerospace*	0.708
Materials Science, Characterization & Testing	0.713	**Limnology**	0.620
Optics	0.690	**Engineering, Petroleum**	0.599
Instruments & Instrumentation	0.666	**Geography**	0.597
Mechanics	0.629	**Environmental Sciences**	0.570
Engineering, Chemical	0.621	**Water Resources**	0.543
Engineering, Electrical & Electronic	0.583	**Engineering, Civil**	0.526
Chemistry, Inorganic & Nuclear	0.566	**Imaging Science & Photographic Technology**	0.514
Crystallography	0.563		
Spectroscopy	0.562		
Chemistry, Multidisciplinary	0.554		
Materials Science, Composites	0.537		
Engineering	0.513		

Factor 5	% variance	Factor 6	% variance
Management, Law & Economy	4.7	Computer Science & Telecommunications	3.8
Planning & Development	0.861	*Computer Science, Artificial Intelligence*	0.922
Industrial Relations & Labor	0.843	*Computer Science, Hardware & Architecture*	0.888
Area Studies	0.842	*Computer Science, Information Systems*	0.848
Public Administration	0.837	*Computer Science, Theory & Methods*	0.832
Urban Studies	0.802	*Telecommunications*	0.832
Political Science	0.766	*Computer Science, Software, Graphics, Programming*	0.824
Law	0.764	*Automation & Control Systems*	0.789
Agricultural Economics & Policy	0.758	**Computer Science, Cybernetics**	0.691
International Relations	0.757		
History Of Social Sciences	0.727		
Environmental Studies	0.713		
Business, Finance	0.697		
Ethnic Studies	0.656		
History	0.619		
Architecture	0.601		
Sociology	0.593		
Social Sigues	0.585		
Demography	0.531		
Communication	0.521		

Factor 7	% variance	Factor 8	% variance
Animal Biology & Ecology	3	Humanities	2.8
Ornithology	0.810	*Language & Linguistics*	0.859
Entomology	0.687	*Literature, American*	0.850
Ecology	0.643	*Literary Reviews*	0.816
Anthropology	0.604	*Poetry*	0.815
Marine & Freshwater Biology	0.598	**Theater**	0.709
Fisheries	0.578	**Art**	0.686
Zoology	0.570	**Arts & Humanities, General**	0.650

Factor 9	% variance	Factor 10	% variance
Health Policy & Services	2	Agriculture & Soil Sciences	2
Health Policy & Services	0.802	*Horticulture*	0.735
Nursing	0.781	*Agriculture*	0.704
Social Sciences, Biomedical	0.693	**Forestry**	0.529
Health Care Sciences & Services	0.678	**Agriculture, Soil Science**	0.508
Medical Informatics	0.632		
Education, Scientific Disciplines	0.593		
Emergency Medicine & Critical Care	0.532		
Tropical Medicine	0.501		

Factor 11	% variance	Factor 12	% variance
Engineering Mechanical	1.9	Orthopedics	1.6
Engineering, Mechanical	0.710	*Orthopedics*	0.704
Engineering	0.661	**Otorhinolaryngology**	0.571
Thermodynamics	0.651		
Materials Science, Composites	0.625		
Materials Science, Characterization & Testing	0.585		
Construction & Building Technology	0.561		
Engineering, Marine	0.522		
Acoustics	0.513		

Factor 13	% variance	Factor 14	% variance
Applied Mathematics	1.4	Physics, Particles & Fields	1.1
Social Sciences, Mathematical Methods	0.735	*Physics, Nuclear*	0.768
Statistics & Probability	0.734	*Physics, Particles & Fields*	0.760
Mathematics, Miscellaneous	0.677	*Physics, Mathematical*	0.713
Mathematics, Applied	0.647	**Physics, Fluids & Plasmas**	0.586
Psychology, Mathematical	0.612	**Nuclear Science & Technology**	0.516
Economics	0.563		
Business, Finance	0.526		

Annex IV

Factors Extracted from the Domain Spain, 1990–1994

Factor 1	% variance	Factor 2	% variance
Biomedicine	23.100	Materials Science & Applied Physics	12.9
Gastroenterology & Hepatology	0.954	*Polymer Science*	0.888
Pediatrics	0.946	*Materials Science, Ceramics*	0.878
Urology & Nephrology	0.930	*Physics, Atomic, Molecular & Chemical*	0.872
Pathology	0.924	*Communication*	0.859
Obstetrics & Gynecology	0.913	*Materials Science, Coatings & Films*	0.855
Oncology	0.905	*Materials Science, Multidisciplinary*	0.851
Medical Laboratory Technology	0.902	*Materials Science, Characterization & Testing*	0.850
Ophthalmology	0.902	*Crystallography*	0.842
Geriatrics & Gerontology	0.901	*Physics, Condensed Matter*	0.838
Hematology	0.901	*Education, Scientific Disciplines*	0.830
Medicine, Research & Experimental	0.895	*Chemistry, Physical*	0.829
Dermatology & Venereal Diseases	0.889	*Thermodynamics*	0.827
Rheumatology	0.882	*Electrochemistry*	0.827
Endocrinology & Metabolism	0.881	*Metallurgy & Metallurgical Engineering*	0.820
Surgery	0.877	*Spectroscopy*	0.813
Anesthesiology	0.873	*Physics, Applied*	0.805
Radiology, Nuclear Medicine & Medical Imaging	0.865	*Chemistry, Inorganic & Nuclear*	0.786
Respiratory System	0.842	*Engineering, Chemical*	0.767
Medicine, General & Internal	0.841	*Physics, Multidisciplinary*	0.765
Cardiac & Cardiovascular Systems	0.837	*Instruments & Instrumentation*	0.744
Sport Sciences	0.836	*Mechanics*	0.743

Factor 1	% variance	Factor 2	% variance
Immunology	0.834	*Optics*	0.738
Clinical Neurology	0.829	*Engineering, Electrical & Electronic*	0.702
Physiology	0.827	**Computer Science, Interdisciplinary Applications**	0.684
Neurosciences	0.813	**Mineralogy**	0.628
Transplantation	0.807	**Nuclear Science & Technology**	0.625
Peripheral Vascular Disease	0.797	**Chemistry, Organic**	0.604
Substance Abuse	0.791	**Chemistry, Multidisciplinary**	0.583
Public, Environmental & Occupational Health	0.787	**Energy & Fuels**	0.581
Emergency Medicine & Critical Care	0.786	**Materials Science, Composites**	0.567
Dentistry, Oral Surgery & Medicine	0.785	**Physics, Fluids & Plasmas**	0.556
Otorhinolaryngology	0.782	**Physics, Mathematical**	0.550
Anatomy & Morphology	0.780	**Engineering, Manufacturing**	0.534
Orthopedics	0.780	**Physics, Nuclear**	0.529
Rehabilitation	0.774		
Nutrition & Dietetics	0.773		
Pharmacology & Pharmacy	0.771		
Allergy	0.758		
Reproductive Systems	0.753		
Toxicology	0.745		
Cell Biology	0.744		
Biology	0.740		
Developmental Biology	0.728		
Engineering, Biomedical	0.724		
Psychiatry	0.717		
Biophysics	0.710		
Andrology	0.699		
Infectious Diseases	0.683		
Veterinary Sciences	0.678		
Virology	0.663		
Biochemistry & Molecular Biology	0.648		
Chemistry, Medicinal	0.640		
Biochemical Research Methods	0.639		
Microscopy	0.631		

Factor 1	% variance	Factor 2	% variance
Genetics & Heredity	0.630		
Microbiology	0.619		
Zoology	0.612		
Medicine, Legal	0.610		
Theater	0.592		
Behavioral Sciences	0.583		
Parasitology	0.562		
Tropical Medicine	0.547		
Chemistry, Analytical	0.535		
Biotechnology & Applied Microbiology	0.509		

Factor 3	% variance	Factor 4	% variance
Psychology	7.8	Agriculture & Soil Sciences	6.3
Psychology, Social	0.934	*Food Science & Technology*	0.822
Psychology, Developmental	0.904	*Agriculture*	0.813
Psychology, Applied	0.882	*Horticulture*	0.806
Psychology, Educational	0.850	*Agriculture, Soil Science*	0.785
Psychology, Experimental	0.837	*Plant Sciences*	0.782
Psychology, Psychoanalysis	0.820	*Chemistry, Applied*	0.772
Criminology & Penology	0.819	*Agriculture, Dairy & Animal Science*	0.767
Social Work	0.818	*Entomology*	0.737
Psychology, Clinical	0.794	*Mycology*	0.728
Education & Educational Research	0.787	*Materials Science, Paper & Wood*	0.723
Psychology, Mathematical	0.781	*Biotechnology & Applied Microbiology*	0.706
Sociology	0.779	**Forestry**	0.670
Women's Studies	0.772	**Environmental Sciences**	0.669
Family Studies	0.768	**Engineering, Environmental**	0.622
Language & Linguistics	0.677	**Biochemical Research Methods**	0.610
Engineering, Industrial	0.658	**Chemistry, Analytical**	0.607
Psychology	0.625	**Microbiology**	0.575
Ergonomics	0.591	**Genetics & Heredity**	0.574
Education, Special	0.554	**Water Resources**	0.539
Social Sciences, Biomedical	0.548	**Biophysics**	0.513
Social Sciences, Interdisciplinary	0.519		
Psychology, Biological	0.519		

Factor 5	% variance	Factor 6	% variance
Computer Science & Telecommunications	4.1	Earth & Space Sciences	3.9
Computer Science, Theory & Methods	0.889	*Geology*	0.884
Computer Science, Artificial Intelligence	0.880	*Meteorology & Atmospheric Sciences*	0.850
Computer Science, Cybernetics	0.853	*Geochemistry & Geophysics*	0.838
Information Science & Library Science	0.846	*Paleontology*	0.824
Computer Science, Software, Graphics, Programming	0.767	*Geography*	0.820
Automation & Control Systems	0.762	*Geosciences, Interdisciplinary*	0.740
Computer Science, Hardware & Architecture	0.750	*Engineering, Petroleum*	0.732
Computer Science, Information Systems	0.730	*Astronomy & Astrophysics*	0.720
Operations Research & Management Science	0.713	*Mining & Mineral Processing*	0.705
Telecommunications	0.629	**Remote Sensing**	0.695
Ergonomics	0.562	**Engineering, Marine**	0.680
Management	0.535	**Engineering, Aerospace**	0.670
Engineering, Manufacturing	0.534	**Imaging Science & Photographic Technology**	0.594
Medical Informatics	0.531	**Mineralogy**	0.569
		Oceanography	0.523

Factor 7	% variance	Factor 8	% variance
Management, Law & Economy	3.3	Arts & Humanities	2.4
Business, Finance	0.899	*Literature*	0.910
Industrial Relations & Labor	0.898	*Folklore*	0.890
Agricultural Economics & Policy	0.828	*Literature, Romance*	0.820
Business	0.783	*Music*	0.785
Planning & Development	0.775	*Classics*	0.776
Social Sciences, Mathematical Methods	0.768	*Literary Reviews*	0.762
Law	0.768	*History*	0.751
International Relations	0.750	*Art*	0.750
Public Administration	0.725	**Literature, British Isles**	0.698
Political Science	0.698	**Religion**	0.587
Economics	0.694		
History Of Social Sciences	0.667		
Urban Studies	0.662		
Transportation	0.575		
Management	0.542		

Factor 9	% variance	Factor 10	% variance
Animal Biology, Ecology	2.4	Physics, Particles & Fields	1.5
Ornithology	0.839	*Physics, Particles & Fields*	0.743
Literature, African, Australian, Canadian	0.717	**Physics, Nuclear**	0.693
Fisheries	0.675	**Physics, Mathematical**	0.690
Marine & Freshwater Biology	0.663	**Physics, Fluids & Plasmas**	0.576
Ecology	0.637	**Mathematics, Applied**	0.514
Limnology	0.589	**Optics**	0.509
Oceanography	0.578		
Biology, Miscellaneous	0.553		

Factors Extracted from the Domain Spain, 1995–1998

Factor 1	% variance	Factor 2	% variance
Biomedicine	23.1	Materials Science & Applied Physics	13.2
Gastroenterology & Hepatology	0.954	*Crystallography*	0.892
Pathology	0.941	*Materials Science, Ceramics*	0.873
Pediatrics	0.932	*Physics, Atomic, Molecular & Chemical*	0.871
Oncology	0.931	*Electrochemistry*	0.855
Urology & Nephrology	0.923	*Materials Science, Coatings & Films*	0.850
Medical Laboratory Technology	0.922	*Chemistry, Physical*	0.838
Medicine, Research & Experimental	0.921	*Chemistry, Inorganic & Nuclear*	0.835
Ophthalmology	0.919	*Education, Scientific Disciplines*	0.833
Rheumatology	0.918	*Materials Science, Multidisciplinary*	0.831
Hematology	0.916	*Physics, Condensed Matter*	0.824
Obstetrics & Gynecology	0.916	*Spectroscopy*	0.808
Dermatology & Venereal Diseases	0.908	*Thermodynamics*	0.808
Endocrinology & Metabolism	0.905	*Metallurgy & Metallurgical Engineering*	0.795
Geriatrics & Gerontology	0.897	*Communication*	0.794
Immunology	0.873	*Engineering, Chemical*	0.783

Factor 1	% variance	Factor 2	% variance
Surgery	0.869	*Physics, Applied*	0.771
Physiology	0.863	*Materials Science, Characterization & Testing*	0.771
Medicine, General & Internal	0.856	*Physics, Multidisciplinary*	0.727
Anesthesiology	0.856	*Optics*	0.712
Cardiac & Cardiovascular Systems	0.844	**Chemistry, Organic**	0.693
Transplantation	0.841	**Instruments & Instrumentation**	0.676
Respiratory System	0.839	**Mechanics**	0.670
Radiology, Nuclear Medicine & Medical Imaging	0.834	**Chemistry, Multidisciplinary**	0.661
Neurosciences	0.832	**Engineering, Electrical & Electronic**	0.613
Pharmacology & Pharmacy	0.831	**Mineralogy**	0.608
Sport Sciences	0.828	**Computer Science, Interdisciplinary Applications**	0.600
Public, Environmental & Occupational Health	0.822	**Materials Science, Composites**	0.590
Peripheral Vascular Disease	0.820	**Physics, Fluids & Plasmas**	0.575
Dentistry, Oral Surgery & Medicine	0.819	**Materials Science, Textiles**	0.570
Anatomy & Morphology	0.809	**Nuclear Science & Technology**	0.563
Emergency Medicine & Critical Care	0.806	**Energy & Fuels**	0.561
Clinical Neurology	0.804	**Physics, Mathematical**	0.505
Cell Biology	0.791		
Otorhinolaryngology	0.782		
Allergy	0.780		
Toxicology	0.777		
Reproductive Systems	0.772		
Developmental Biology	0.771		
Biophysics	0.760		
Virology	0.756		
Substance Abuse	0.749		
Biology	0.743		
Infectious Diseases	0.733		
Andrology	0.727		
Nutrition & Dietetics	0.726		

Factor 1	% variance	Factor 2	% variance
Veterinary Sciences	0.723		
Genetics & Heredity	0.721		
Engineering, Biomedical	0.718		
Microscopy	0.714		
Orthopedics	0.706		
Chemistry, Medicinal	0.701		
Biochemistry & Molecular Biology	0.697		
Rehabilitation	0.695		
Microbiology	0.688		
Biochemical Research Methods	0.687		
Medicine, Legal	0.683		
Psychiatry	0.681		
Tropical Medicine	0.611		
Parasitology	0.600		
Biotechnology & Applied Microbiology	0.593		
Zoology	0.582		
Behavioral Sciences	0.577		
Chemistry, Analytical	0.564		
Acoustics	0.512		
Mycology	0.502		

Factor 3	% variance	Factor 4	% variance
Psychology	8.2	Agriculture & Soil Sciences	6.8
Psychology, Developmental	0.909	*Food Science & Technology*	0.826
Psychology, Applied	0.874	*Agriculture, Dairy & Animal Science*	0.823
Criminology & Penology	0.865	*Agriculture*	0.816
Psychology, Educational	0.863	*Agriculture, Soil Science*	0.800
Psychology, Psychoanalysis	0.845	*Materials Science, Paper & Wood*	0.793
Psychology, Experimental	0.826	*Chemistry, Applied*	0.775
Social Work	0.821	*Entomology*	0.757
Education & Educational Research	0.813	*Plant Sciences*	0.754
Family Studies	0.812	*Forestry*	0.714
Psychology, Clinical	0.807	**Agricultural Economics & Policy**	0.677
Education, Special	0.758	**Mycology**	0.648

Factor 3	% variance	Factor 4	% variance
Sociology	0.729	**Biotechnology & Applied Microbiology**	0.630
Language & Linguistics	0.719	**Environmental Sciences**	0.627
Psychology	0.694	**Engineering, Environmental**	0.614
Women's Studies	0.688	**Film, Radio, Television**	0.565
Psychology, Mathematical	0.678	**Chemistry, Analytical**	0.556
Psychology, Biological	0.575	**Nutrition & Dietetics**	0.538
Rehabilitation	0.518	**Music**	0.537
Psychiatry	0.516	**Biochemical Research Methods**	0.521
Substance Abuse	0.513	**Water Resources**	0.520

Factor 5	% variance	Factor 6	% variance
Computer Science & Telecommunications	4.4	Earth & Space Sciences	4
Computer Science, Cybernetics	0.894	*Meteorology & Atmospheric Sciences*	0.877
Computer Science, Artificial Intelligence	0.873	*Geochemistry & Geophysics*	0.858
Computer Science, Hardware & Architecture	0.871	*Paleontology*	0.780
Computer Science, Software, Graphics, Programming	0.869	*Astronomy & Astrophysics*	0.779
Information Science & Library Science	0.809	*Geosciences, Interdisciplinary*	0.755
Automation & Control Systems	0.799	*Geography*	0.736
Computer Science, Information Systems	0.779	*Engineering, Petroleum*	0.713
Ergonomics	0.730	*Remote Sensing*	0.709
Operations Research & Management Science	0.711	*Engineering, Geological*	0.702
Transportation	0.635	**Engineering, Marine**	0.677
Telecommunications	0.622	**Oceanography**	0.671
Management	0.568	**Mining & Mineral Processing**	0.671
Engineering, Industrial	0.562	**Imaging Science & Photographic Technology**	0.641
		Engineering, Aerospace	0.619
		Mineralogy	0.597

Factor 5	% variance	Factor 6	% variance
		Limnology	0.595
		Water Resources	0.588
		Literature, American	0.561
		Engineering, Civil	0.518

Factor 7	% variance	Factor 8	% variance
Management, Law & Economy	3.4	Humanities	3.1
Industrial Relations & Labor	0.908	*Literature, African, Australian, Canadian*	0.801
Planning & Development	0.873	*History*	0.791
Business	0.868	*Literature, Romance*	0.775
Political Science	0.864	*Literary Reviews*	0.766
Law	0.779	*Folklore*	0.710
Social Sciences, Mathematical Methods	0.770	*Literature*	0.702
International Relations	0.729	**Theater**	0.681
Urban Studies	0.699	**Asian Studies**	0.520
Economics	0.692		
Public Administration	0.691		
Area Studies	0.592		
Management	0.588		
Agricultural Economics & Policy	0.549		
History Of Social Sciences	0.520		
Environmental Studies	0.507		

Factor 9	% variance	Factor 10	% variance
Animal Biology, Ecology	2.3	Physics, Particles & Fields	1.4
Anthropology	0.708	*Physics, Nuclear*	0.731
Arts & Humanities, General	0.685	*Physics, Mathematical*	0.711
Ecology	0.641	**Physics, Fluids & Plasmas**	0.638
Biology, Miscellaneous	0.616	**Nuclear Science & Technology**	0.561
Zoology	0.549	**Communication**	0.543
Marine & Freshwater Biology	0.548	**Optics**	0.541
Fisheries	0.547		
Geography	0.503		

Factors Extracted from the Domain Spain, 1999–2002

Factor 1	% variance	Factor 2	% variance
Biomedicine	22.1	Materials Science & Applied Physics	12.3
Gastroenterology & Hepatology	0.952	*Materials Science, Ceramics*	0.939
Pathology	0.933	*Metallurgy & Metallurgical Engineering*	0.914
Urology & Nephrology	0.927	*Materials Science, Characterization & Testing*	0.895
Pediatrics	0.926	*Materials Science, Coatings & Films*	0.893
Rheumatology	0.921	*Polymer Science*	0.865
Medicine, Research & Experimental	0.911	*Physics, Condensed Matter*	0.851
Dermatology & Venereal Diseases	0.907	*Electrochemistry*	0.838
Medical Laboratory Technology	0.906	*Materials Science, Composites*	0.820
Oncology	0.902	*Physics, Atomic, Molecular & Chemical*	0.819
Hematology	0.897	*Physics, Applied*	0.809
Endocrinology & Metabolism	0.887	*Materials Science, Multidisciplinary*	0.808
Obstetrics & Gynecology	0.871	*Crystallography*	0.797
Surgery	0.868	*Chemistry, Physical*	0.779
Immunology	0.865	*Thermodynamics*	0.750
Ophthalmology	0.857	*Communication*	0.733
Geriatrics & Gerontology	0.854	*Mechanics*	0.714
Medicine, General & Internal	0.854	*Engineering, Chemical*	0.712
Physiology	0.850	*Chemistry, Inorganic & Nuclear*	0.712
Anesthesiology	0.830	*Physics, Multidisciplinary*	0.711
Transplantation	0.825	*Education, Scientific Disciplines*	0.703
Pharmacology & Pharmacy	0.824	**Optics**	0.694
Cardiac & Cardiovascular Systems	0.821	**Spectroscopy**	0.657
Respiratory System	0.821	**Construction & Building Technology**	0.631
Dentistry, Oral Surgery & Medicine	0.816	**Instruments & Instrumentation**	0.631
Public, Environmental & Occupational Health	0.810	**Energy & Fuels**	0.623

Factor 1	% variance	Factor 2	% variance
Sport Sciences	0.805	**Engineering, Electrical & Electronic**	0.617
Neurosciences	0.804	**Chemistry, Multidisciplinary**	0.610
Radiology, Nuclear Medicine & Medical Imaging	0.799	**Mineralogy**	0.597
Peripheral Vascular Disease	0.798	**Engineering, Manufacturing**	0.588
Otorhinolaryngology	0.797	**Physics, Fluids & Plasmas**	0.579
Allergy	0.776	**Chemistry, Organic**	0.550
Cell Biology	0.768	**Materials Science, Biomaterials**	0.543
Virology	0.763	**Mining & Mineral Processing**	0.540
Anatomy & Morphology	0.759	**Engineering, Mechanical**	0.511
Clinical Neurology	0.758		
Biophysics	0.743		
Developmental Biology	0.732		
Infectious Diseases	0.727		
Toxicology	0.724		
Orthopedics	0.723		
Reproductive Systems	0.716		
Microscopy	0.705		
Emergency Medicine & Critical Care	0.705		
Genetics & Heredity	0.696		
Critical Care Medicine	0.688		
Biology	0.687		
Poetry	0.684		
Rehabilitation	0.681		
Microbiology	0.678		
Biochemistry & Molecular Biology	0.675		
Engineering, Biomedical	0.672		
Veterinary Sciences	0.671		
Substance Abuse	0.669		
Film, Radio, Television	0.660		
Chemistry, Medicinal	0.659		
Nutrition & Dietetics	0.650		
Andrology	0.641		
Biochemical Research Methods	0.636		
Psychiatry	0.632		
Neuroimaging	0.632		
Literature, American	0.622		

Factor 1	% variance	Factor 2	% variance
Gerontology	0.611		
Medicine, Legal	0.595		
Tropical Medicine	0.587		
Biotechnology & Applied Microbiology	0.578		
Parasitology	0.572		
Health Care Sciences & Services	0.518		
Behavioral Sciences	0.510		
Integrative & Complementary Medicine	0.504		

Factor 3	% variance	Factor 4	% variance
Psychology	7.7	Earth & Space Sciences	6.9
Psychology, Social	0.934	*Geology*	0.883
Psychology, Developmental	0.884	*Meteorology & Atmospheric Sciences*	0.878
Psychology, Applied	0.880	*Geochemistry & Geophysics*	0.857
Psychology, Educational	0.869	*Astronomy & Astrophysics*	0.807
Psychology, Experimental	0.866	*Paleontology*	0.773
Psychology, Multidisciplinary	0.849	*Geosciences, Interdisciplinary*	0.772
Education & Educational Research	0.844	*Geography, Physical*	0.771
Criminology & Penology	0.824	*Oceanography*	0.762
Psychology, Psychoanalysis	0.823	*Engineering, Ocean*	0.760
Psychology, Clinical	0.790	*Remote Sensing*	0.735
Social Work	0.767	*Engineering, Geological*	0.723
Psychology, Mathematical	0.762	*Engineering, Petroleum*	0.710
Family Studies	0.748	**Engineering, Marine**	0.699
Psychology	0.729	**Engineering, Aerospace**	0.678
Sociology	0.705	**Geography**	0.667
Education, Special	0.693	**Imaging Science & Photographic Technology**	0.656
Applied Linguistics	0.677	**Limnology**	0.602
Language & Linguistics	0.669	**Mining & Mineral Processing**	0.588
Psychology, Biological	0.639	**Water Resources**	0.569
Women's Studies	0.571	**Mineralogy**	0.562
Psychiatry	0.525		
Substance Abuse	0.517		

Factor 5	% variance	Factor 6	% variance
Computer Science & Telecommunications	4.4	Agriculture & Soil Sciences	4.1
Computer Science, Cybernetics	0.903	*Agriculture, Multidisciplinary*	0.877
Computer Science, Hardware & Architecture	0.886	*Food Science & Technology*	0.848
Computer Science, Artificial Intelligence	0.874	*Horticulture*	0.843
Computer Science, Theory & Methods	0.871	*Agricultural Engineering*	0.820
Computer Science, Software, Graphics, Programming	0.855	*Materials Science, Paper & Wood*	0.807
Automation & Control Systems	0.822	*Chemistry, Applied*	0.800
Ergonomics	0.800	*Agriculture*	0.798
Robotics	0.771	*Agriculture, Dairy & Animal Science*	0.793
Computer Science, Information Systems	0.767	*Agriculture, Soil Science*	0.773
Information Science & Library Science	0.755	**Plant Sciences**	0.691
Telecommunications	0.710	**Entomology**	0.671
Operations Research & Management Science	0.610	**Materials Science, Textiles**	0.641
		Nutrition & Dietetics	0.625
		Chemistry, Analytical	0.601
		Environmental Sciences	0.592
		Biotechnology & Applied Microbiology	0.582
		Engineering, Environmental	0.577
		Forestry	0.567
		Mycology	0.551
		Biochemical Research Methods	0.540
		Water Resources	0.509

Factor 7	% variance	Factor 8	% variance
Management, Law & Economy	3.4	Animal Biology, Ecology	2.7
Industrial Relations & Labor	0.936	*Biodiversity Conservation*	0.909
Business, Finance	0.933	*Ornithology*	0.895
Political Science	0.886	*Evolutionary Biology*	0.849
International Relations	0.875	*Anthropology*	0.779
Planning & Development	0.873	*Ecology*	0.757
Law	0.845	*Biology, Miscellaneous*	0.749

Factor 7	% variance	Factor 8	% variance
Business	0.827	*Arts & Humanities, General*	0.727
Social Sciences, Mathematical Methods	0.779	**Zoology**	0.690
Public Administration	0.739	**Entomology**	0.627
Agricultural Economics & Policy	0.710	**Marine & Freshwater Biology**	0.621
		Forestry	0.594
		Fisheries	0.588
		Social Sciences, Interdisciplinary	0.546
		Archeology	0.539
		Geography	0.511

Factor 9	% variance	Factor 10	% variance
Health Policy & Services	2.1	Humanities	1.9
Social Sciences, Biomedical	0.726	*Literature, Romance*	0.813
Medical Ethics	0.719	*Classics*	0.807
Social Issues	0.713	*Literature*	0.803
Health Policy & Services	0.697	**Religion**	0.681
Ethics	0.683	**Literary Reviews**	0.597
Nursing	0.644	**Asian Studies**	0.581
Demography	0.622	**Area Studies**	0.501
Health Care Sciences & Services	0.592		
Women's Studies	0.571		
Philosophy	0.533		

Factor 11	% variance
Physics, Particles & Fields	1.7
Physics, Particles & Fields	0.780
Physics, Nuclear	0.771
Physics, Mathematical	0.708
Communication	0.613
Nuclear Science & Technology	0.604
Physics, Fluids & Plasmas	0.602
Optics	0.559

Annex V

ANEP Classes

Note Form	ANEP Classes
AGR	Agricultura
MOL	Biologia Molecular, Celular Y Genetica
VEG	Biologia Vegetal Y Animal, Ecologia
ALI	Ciencia Y Tecnologia De Alimentos
MAR	Ciencia Y Tecnologia De Materiales
COM	Ciencias De La Computacion Y Tecnologia Informatica
TIE	Ciencias De La Tierra
CSS	Ciencias Sociales
DER	Derecho
ECO	Economia
FIL	Filologia Y Filosofia
FIS	Fisica Y Ciencias Del Espacio
FAR	Fisiologia Y Farmacologia
GAN	Ganaderia Y Pesca
HIS	Historia Y Arte
CIV	Ingenieria Civil Y Arquitectura
ELE	Ingenieria Electrica, Electronica Y Automatica
MEC	Ingenieria Mecanica, Naval Y Aeronautica
MAT	Matematicas
MED	Medicina
PSI	Psicologia Y Ciencias De La Educacion
QUI	Quimica
TEC	Tecnologia Electronica Y De Las Comunicaciones
TQU	Tecnologia Quimica

Correspondence Between ANEP Classes and JCR Categories

ANEP Classes	JCR Categories
Agriculture	Agriculture, Dairy & Animal Science
Agriculture	Agriculture
Agriculture	Agriculture, Soil Science
Agriculture	Biotechnology & Applied Microbiology
Agriculture	Environmental Sciences
Agriculture	Agricultural Economics & Policy
Agriculture	Forestry
Agriculture	Horticulture
Agriculture	Agriculture, Multidisciplinary
Agriculture	Agricultural Engineering
Chemical Technology	Metallurgy & Metallurgical Engineering
Chemical Technology	Engineering, Chemical
Chemical Technology	Materials Science, Paper & Wood
Chemical Technology	Mining & Mineral Processing
Chemical Technology	Materials Science, Textiles
Chemistry	Education, Scientific Disciplines
Chemistry	Chemistry, Multidisciplinary
Chemistry	Environmental Sciences
Chemistry	Chemistry, Organic
Chemistry	Chemistry, Analytical
Chemistry	Chemistry, Physical
Chemistry	Chemistry, Inorganic & Nuclear
Chemistry	Toxicology
Chemistry	Engineering, Environmental
Chemistry	Chemistry, Applied
Chemistry	Electrochemistry
Civil Engineering & Architecture	Construction & Building Technology
Civil Engineering & Architecture	Engineering
Civil Engineering & Architecture	Computer Science, Interdisciplinary Applications
Civil Engineering & Architecture	Engineering, Civil
Civil Engineering & Architecture	Mining & Mineral Processing
Civil Engineering & Architecture	Transportation
Computer Science & Technology	Computer Science, Theory & Methods
Computer Science & Technology	Computer Science, Information Systems

ANEP Classes	JCR Categories
Computer Science & Technology	Computer Science, Software, Graphics, Programming
Computer Science & Technology	Automation & Control Systems
Computer Science & Technology	Computer Science, Artificial Intelligence
Computer Science & Technology	Computer Science, Hardware & Architecture
Computer Science & Technology	Computer Science, Interdisciplinary Applications
Computer Science & Technology	Computer Science, Cybernetics
Computer Science & Technology	Robotics
Economy	Economics
Economy	Business
Economy	Management
Economy	Business, Finance
Electrical, Electronic & Automated Engineering	Engineering, Electrical & Electronic
Electrical, Electronic & Automated Engineering	Engineering
Electrical, Electronic & Automated Engineering	Remote Sensing
Electrical, Electronic & Automated Engineering	Robotics
Electronic & Telecommunications Technology	Engineering, Electrical & Electronic
Electronic & Telecommunications Technology	Telecommunications
Electronic & Telecommunications Technology	Computer Science, Hardware & Architecture
Electronic & Telecommunications Technology	Imaging Science & Photographic Technology
Food Science And Technology	Nutrition & Dietetics
Food Science And Technology	Food Science & Technology
Food Science And Technology	Biotechnology & Applied Microbiology
Geosciences	Energy & Fuels
Geosciences	Geosciences, Interdisciplinary
Geosciences	Engineering, Petroleum
Geosciences	Crystallography
Geosciences	Environmental Sciences
Geosciences	Water Resources
Geosciences	Meteorology & Atmospheric Sciences
Geosciences	Paleontology
Geosciences	Geochemistry & Geophysics
Geosciences	Mineralogy
Geosciences	Engineering, Environmental

ANEP Classes	JCR Categories
Geosciences	Oceanography
Geosciences	Geography
Geosciences	Geology
Geosciences	Imaging Science & Photographic Technology
Geosciences	Engineering, Geological
Geosciences	Engineering, Ocean
Geosciences	Geography, Physical
History & Arts	History & Philosophy Of Science
History & Arts	Urban Studies
History & Arts	Archeology
History & Arts	History
History & Arts	Music
History & Arts	Architecture
History & Arts	Art
History & Arts	Folklore
History & Arts	Film, Radio, Television
Law	Law
Law	Criminology & Penology
Law	International Relations
Livestock & Fishing	Agriculture, Dairy & Animal Science
Livestock & Fishing	Biotechnology & Applied Microbiology
Livestock & Fishing	Veterinary Sciences
Livestock & Fishing	Fisheries
Materials Science and Technology	Materials Science, Multidisciplinary
Materials Science and Technology	Crystallography
Materials Science and Technology	Polymer Science
Materials Science and Technology	Materials Science, Composites
Materials Science and Technology	Materials Science, Ceramics
Materials Science and Technology	Materials Science, Coatings & Films
Materials Science and Technology	Materials Science, Biomaterials
Materials Science and Technology	Materials Science, Characterization & Testing
Mathematics	Mathematics
Mathematics	Mathematics, Applied
Mathematics	Statistics & Probability

ANEP Classes	JCR Categories
Mathematics	Automation & Control Systems
Mathematics	Operations Research & Management Science
Mathematics	Mathematics, Miscellaneous
Mechanical, Naval & Aeronautic Engineering	Engineering, Aerospace
Mechanical, Naval & Aeronautic Engineering	Mechanics
Mechanical, Naval & Aeronautic Engineering	Engineering, Mechanical
Mechanical, Naval & Aeronautic Engineering	Engineering
Mechanical, Naval & Aeronautic Engineering	Engineering, Manufacturing
Mechanical, Naval & Aeronautic Engineering	Engineering, Marine
Mechanical, Naval & Aeronautic Engineering	Engineering, Industrial
Medicine	Gastroenterology & Hepatology
Medicine	Radiology, Nuclear Medicine & Medical Imaging
Medicine	Medicine, General & Internal
Medicine	Medical Informatics
Medicine	Cardiac & Cardiovascular Systems
Medicine	Respiratory System
Medicine	Nutrition & Dietetics
Medicine	Anesthesiology
Medicine	Anatomy & Morphology
Medicine	Biology
Medicine	Surgery
Medicine	Dermatology & Venereal Diseases
Medicine	Endocrinology & Metabolism
Medicine	Hematology
Medicine	Medicine, Research & Experimental
Medicine	Neurosciences
Medicine	Clinical Neurology
Medicine	Pathology
Medicine	Obstetrics & Gynecology
Medicine	Dentistry, Oral Surgery & Medicine
Medicine	Oncology
Medicine	Ophthalmology
Medicine	Orthopedics
Medicine	Otorhinolaryngology

ANEP Classes	JCR Categories
Medicine	Pediatrics
Medicine	Psychiatry
Medicine	Tropical Medicine
Medicine	Parasitology
Medicine	Substance Abuse
Medicine	Medical Laboratory Technology
Medicine	Toxicology
Medicine	Geriatrics & Gerontology
Medicine	Infectious Diseases
Medicine	Rheumatology
Medicine	Urology & Nephrology
Medicine	Allergy
Medicine	Public, Environmental & Occupational Health
Medicine	Emergency Medicine & Critical Care
Medicine	Medicine, Legal
Medicine	Peripheral Vascular Disease
Medicine	Rehabilitation
Medicine	Sport Sciences
Medicine	Andrology
Medicine	Engineering, Biomedical
Medicine	Transplantation
Medicine	Social Sciences, Biomedical
Medicine	Nursing
Medicine	Health Policy & Services
Medicine	Health Care Sciences & Services
Medicine	Integrative & Complementary Medicine
Medicine	Critical Care Medicine
Medicine	Medical Ethics
Medicine	Neuroimaging
Molecular & Cellular Biology & Genetics	Anatomy & Morphology
Molecular & Cellular Biology & Genetics	Biochemistry & Molecular Biology
Molecular & Cellular Biology & Genetics	Biology
Molecular & Cellular Biology & Genetics	Biotechnology & Applied Microbiology
Molecular & Cellular Biology & Genetics	Biochemical Research Methods
Molecular & Cellular Biology & Genetics	Cell Biology

ANEP Classes	JCR Categories
Molecular & Cellular Biology & Genetics	Virology
Molecular & Cellular Biology & Genetics	Microbiology
Molecular & Cellular Biology & Genetics	Biophysics
Molecular & Cellular Biology & Genetics	Genetics & Heredity
Molecular & Cellular Biology & Genetics	Immunology
Molecular & Cellular Biology & Genetics	Toxicology
Molecular & Cellular Biology & Genetics	Biology, Miscellaneous
Molecular & Cellular Biology & Genetics	Developmental Biology
Molecular & Cellular Biology & Genetics	Microscopy
Molecular & Cellular Biology & Genetics	Evolutionary Biology
Philology & Philosophy	History & Philosophy Of Science
Philology & Philosophy	Language & Linguistics
Philology & Philosophy	Philosophy
Philology & Philosophy	Literature, Romance
Philology & Philosophy	Theater
Philology & Philosophy	Literary Reviews
Philology & Philosophy	Literature
Philology & Philosophy	Poetry
Philology & Philosophy	Literature, American
Philology & Philosophy	Arts & Humanities, General
Philology & Philosophy	Religion
Philology & Philosophy	Classics
Philology & Philosophy	Asian Studies
Philology & Philosophy	Literature, German, Netherlandic, Scandinavian
Philology & Philosophy	Literature, British Isles
Philology & Philosophy	Literature, African, Australian, Canadian
Philology & Philosophy	Ethics
Philology & Philosophy	Applied Linguistics
Philology & Philosophy	Literary Theory & Criticism
Physics & Space Sciences	Acoustics
Physics & Space Sciences	Astronomy & Astrophysics
Physics & Space Sciences	Physics, Multidisciplinary

ANEP Classes	JCR Categories
Physics & Space Sciences	Physics, Applied
Physics & Space Sciences	Physics, Nuclear
Physics & Space Sciences	Optics
Physics & Space Sciences	Physics, Atomic, Molecular & Chemical
Physics & Space Sciences	Physics, Condensed Matter
Physics & Space Sciences	Instruments & Instrumentation
Physics & Space Sciences	Nuclear Science & Technology
Physics & Space Sciences	Physics, Fluids & Plasmas
Physics & Space Sciences	Physics, Particles & Fields
Physics & Space Sciences	Spectroscopy
Physics & Space Sciences	Thermodynamics
Physics & Space Sciences	Computer Science, Interdisciplinary Applications
Physics & Space Sciences	Physics, Mathematical
Physiology & Pharmacology	Nutrition & Dietetics
Physiology & Pharmacology	Pharmacology & Pharmacy
Physiology & Pharmacology	Physiology
Physiology & Pharmacology	Substance Abuse
Physiology & Pharmacology	Reproductive Systems
Physiology & Pharmacology	Behavioral Sciences
Physiology & Pharmacology	Chemistry, Medicinal
Physiology & Pharmacology	Psychology, Experimental
Physiology & Pharmacology	Psychology, Biological
Plant & Animal Biology, Ecology	Entomology
Plant & Animal Biology, Ecology	Plant Sciences
Plant & Animal Biology, Ecology	Zoology
Plant & Animal Biology, Ecology	Biology
Plant & Animal Biology, Ecology	Marine & Freshwater Biology
Plant & Animal Biology, Ecology	Ecology
Plant & Animal Biology, Ecology	Veterinary Sciences
Plant & Animal Biology, Ecology	Biology, Miscellaneous
Plant & Animal Biology, Ecology	Limnology
Plant & Animal Biology, Ecology	Ornithology
Plant & Animal Biology, Ecology	Mycology
Plant & Animal Biology, Ecology	Biodiversity Conservation
Plant & Animal Biology, Ecology	Evolutionary Biology
Psychology & Educational Sciences	Behavioral Sciences
Psychology & Educational Sciences	Psychology

ANEP Classes	JCR Categories
Psychology & Educational Sciences	Ergonomics
Psychology & Educational Sciences	Psychology, Mathematical
Psychology & Educational Sciences	Psychology, Developmental
Psychology & Educational Sciences	Education & Educational Research
Psychology & Educational Sciences	Psychology, Educational
Psychology & Educational Sciences	Psychology, Biological
Psychology & Educational Sciences	Psychology, Clinical
Psychology & Educational Sciences	Social Work
Psychology & Educational Sciences	Psychology, Social
Psychology & Educational Sciences	Education, Special
Psychology & Educational Sciences	Psychology, Psychoanalysis
Psychology & Educational Sciences	Psychology, Applied
Psychology & Educational Sciences	Psychology, Multidisciplinary
Social Sciences	History & Philosophy Of Science
Social Sciences	Anthropology
Social Sciences	Information Science & Library Science
Social Sciences	Geography
Social Sciences	Environmental Studies
Social Sciences	Sociology
Social Sciences	Transportation
Social Sciences	Social Sciences, Mathematical Methods
Social Sciences	Business
Social Sciences	Management
Social Sciences	Social Sciences, Interdisciplinary
Social Sciences	Planning & Development
Social Sciences	Social Issues
Social Sciences	Public Administration
Social Sciences	Social Work
Social Sciences	Women's Studies
Social Sciences	Area Studies

ANEP Classes	JCR Categories
Social Sciences	Political Science
Social Sciences	International Relations
Social Sciences	Family Studies
Social Sciences	History Of Social Sciences
Social Sciences	Ethnic Studies
Social Sciences	Industrial Relations & Labor
Social Sciences	Communication
Social Sciences	Demography
Social Sciences	Gerontology

Bibliography

ANEP. (2005). *Agencia Nacional de Evaluación y Prospectiva.* Available from: <http://www.mcyt.es/sepct/ANEP/anep.htm > (Visited: 31/3/2005).

Araya, A.A. (2003). The hidden side of visualization. *Journal of the Society for Philosophy and Technology,* 7, 27-93.

Barnes, J. A. (1954). Class and committees in a Norwegian island parish. *Human Relations,* 7, 3-58.

Batagelj, V. and Mrvar, A. (1998). Pajek: program for large network analysis. *Conections,* 21, 47-57.

Batini, C., Nardelli, E., and Tamassia, R. (1986). A layout algorithm for data flow diagrams. *IEEE Transactions. Software Engineering,* SE-12, 539-546.

Bonnevie, E. (2003). A multifaceted portrait of a library and information science Journal: the case of the Journal of Information Science. *Journal of Information Science,* 29, 11-23.

Bordons, M. and Gómez Caridad, I. (1997). La Actividad Científica Española a través de Indicadores Bibliométricos en el Período 1990-93. *Revista General de Información y Documentación,* 7, (2), 69-86.

Borgatti, S. P. and Everett, M. G. (1992). Regular Blockmodels of Multiway, Multimode Matrices. *Social Networks,* 14, (1-2), 91-120.

Brandenburg, F.J., Himsolt, M., and Rohrer, C. (1995). An Experimental Comparison of Force-Directed and Randomized Graph Drawing Algorithms. *Lecture Notes in Computer Science,* 1027, 87.

Braun, T.; Glanzel, W.; y Schubert, A. (2000). How balanced is the Science Citation Index's journal coverage? a preliminary overview of macrolevel statistical data. In B. Cronin and H. B. Atkins (Eds.). The web of knowledge: a festschrift in honor of Eugene Garfield. New Jersey: Information Today.

Bush, V. (1945). As we may think. *The Atlantic Montly,* 176, 101-108.

Buzydlowski, J. (2002). A Comparison of Self-Organizing Maps and Pathfinder Networks for the Mapping of Co-Cited Authors. Phd. Thesis. Universidad de Drexel.

Buzydlowski, J., White, H. D., and Lin, X. (2002). Term co-occurrence analysis as an interface for digital libraries. *Lecture Notes in Computer Science Series,* 2539, 133-144.

Börner, K., Chen, C., and Boyack, K. W. (2003). Visualizing knowledge domains. *Annual Review of Information Science & Technology,* 37, 179-255.

Carpano, M. (1980). Automatic display of hierarchized graphs for computer-aided decision analysis. *IEEE Transaction on Systems Man and Cibernetics,* SMC-10, 705-715.

Cassi, L. (2003). Information, knowledge and social networks: is a new buzzword coming up? In: DRUID PhD Conference, (20 p.).

Chen, C. (1998a). Bridging the gap: the use of pathfinder networks in visual navigation. *Journal of Visual Languages and Computing*, 9, 267-286.

Chen, C. (1998b). Generalised Similarity Analysis and Pathfinder Network Scaling. *Interacting with computers*, 10, 107-128.

Chen, C. (1999a). Information Visualization and Virtual Environments. Berlin: Springer.

Chen, C. (1999b). Visualising semantic spaces and author co-citation networks in digital libraries. *Information Processing & Management*, 35, 401-420.

Chen, C. and Carr, L. (1999a). A semantic-centric approach to information visualization. In: Proceedings of the conference on Visualization '99: celebrating ten years, San Francisco, CA: IEEE Computer Society Press.

Chen, C. and Carr, L. (1999b). Trailblazing the literature of hypertext: an autohor cocitation analysis (1989-1998). Proceeding of the 10th ACM Conference on Hypertext (Hypertext '99).

Chen, C. and Carr, L. (1999c). Visualizing the evolution of a subject domain: a case study. In: Proceedings of the conference on Visualization '99: celebrating ten years, San Francisco, CA: IEEE Computer Society Press.

Chen, C., Cribbin, T., Macredie, R., and Morar, S. (2002). Visualizing and tracking the growth of competing paradigms: two case studies. *Journal of the American Society for Information Science and Technology (JASIST)*, 53, 678-689.

Chen, C. and Hicks, D. (2004). Tracing knowledge diffusion. *Scientometrics*, 59, 199-211.

Chen, C. and Kuljis, J. (2003). The rising landscape: a visual exploration of superstring revolutions in physics. *Journal of the American Society for Information Science and Technology (JASIST)*, 54, 435-446.

Chen, C. and Paul, R.J. (2001). Visualizing a knowledge domain's intellectual structure. *Computer*, 34, 65-71.

Chen, C., Paul, R. J., and O'keefe, B. (2001). Fitting the jigsaw of citation: information visualization in domain analysis. *Journal of the American Society for Information Science and Technology (JASIST)*, 52, (4), 315-330.

Chen, H., Houston, A.L., Sewell, R.R., and Schatz, B.R. (1998). Internet browsing and searching: user evaluations of category map and concept space techniques. *Journal of the American Society for Information Science (JASIS)*, 49, 582-603.

Chen, H., Ng, T.D., Martinez, J., and Schatz, B.R. (1997). A concept space approach to addressing the vocabulary problem in scientific information retrieval: an experiment on the worm community system. *Journal of the American Society for Information Science (JASIS)*, 48, 17-31.

Chen, H., Nunamaker, J.F.Jr., Orwig, R.E., and Titkova, O. (1998). Information visualization for collaborative computing. *IEEE computer*, 31, 75-82.

Chen, H., Schuffels, C., and Orwig, R.E. (1996). Internet categorization and search: a self-organizing approach. *Journal of visual communication and image representation*, 7, 88-102.

Cohen, J. (1997). Drawing Graphs to Convey Proximity: An Incremental Arrangement Method. *ACM Transactions on Computer-Human Interaction,* 4, 197-229.

Corman, S. R. (1990). Computerized Vs Pencil and Paper Collection of Network Data. *Social Networks,* 12, (4), 375-384.

Costa, J. (1998). La esquemática: visualizar la información. Barcelona: Paidós.

Crosby, A. W. (1997). The Measure of Reality: Quantification and Western Society. London: Cambridge University Press.

Di Battista, G. (1998). Graph Drawing: Algorithms for the Visualization of Graphs. Prentice-Hall.

Ding, Y., Chowdhury, G.G., and Foo, S. (1999). Mapping the intellectual structure of information retrieval studies: an author co-citation analysis, 1987-1997. *Journal of Information Science,* 25, 67-78.

Doreian, P. (1985). Structural equivalence in a psychology journal network. *Journal of the American Society for Information Science (JASIS),* 36, 411-417.

Doreian, P. (1988). Testing structural equivalence hipotheses in a network of geographic journals. *Journal of the American Society for Information Science, (JASIS),* 39, 79-85.

Eades, P. (1984). A Heuristic for Graph Drawing. *Congressus Numerantium,* 42, 149-160.

Encyclopaedia Britannica, Inc. (2005). *Encyclopaedia Britannica Online.* Available from: <http://www.britannica.com/> (Visited: 23/2/2005).

Faba-Pérez, C., Guerrero Bote, V.P., and Moya-Anegón, F.d. (2004). Fundamentos y técnicas cibermétricas: modelos cuantitativos de análisis. Badajoz: Junta de Extremadura, Consejería de Cultura.

Fowler, R. H. y Dearhold, D. W. (1990). Information retrieval using path finder networks. In R. W. Ed. Schvaneveldt (Ed.), Pathfinder associative networks; studies in knowledge organization. Ablex (NJ): Norwood.

Franklin, J. J. y Johnston, R. (1988). Co-citation bibliometric modeling as a tool for S&T policy and R&D management: issues aplication and developments. In A. F. J. van Raan (Ed.), Handbook of quantitative studies of science and technology. Amsterdam: North Holland.

Freeman, L.C. (1979). Centrality in social networks: conceptional clarification. *Social Networks,* 1, 215-239.

Freeman, L. C. (2000a). Social network analysis: definition and history. In A. E. Kazdan (Ed.), Encyclopedia of Psychology. New York: Oxford University Press.

Freeman, L.C. (2000b) Visualizing social networks. *Journal of Social Structure,* 1.

Fruchterman, T. and Reingold, E. (1991). Graph Drawing by Force-Directed Placement. *SoftwarePractice and Experience,* 21, 1129-1164.

García-Guinea J. and Ruis J.D. (1998). The consequences of publishing in journals written in Spanish in Spain. *Interciencia,* 23, 185-187.

Garfield, E. (1976). Social-sciences citation index clusters. *Current contents,* 27, 5-11.

Garfield, E. (1992). Psychology research, 1986-1990: a citationist perspective on the highest impact papers, institutions, and authors. *Current contents,* 41, 5-13.

Garfield, E. (1998). Mapping the world of science. 150 Anniversay Meeting of the AAAS, Philadelphia, PA.

Garfield, E. (1981). Introducing the ISI Atlas of Science: Biochemistry and molecular biology, 1978-80. *Current Contents*, (42), 5-13.

Garfield, E. (1994). Scientography: mapping the tracks of science. *Current contents: social & behavioral sciences*, 7, 5-10.

Garfield, E., Sher, I.H., and Torpie, R.J. (1964). The use of citation data in writing the history of science. Philadelphia: Institute for Scientific Information.

Gosper, Jeffrey J. (1998). *Floyd-Warshall all-pairs shortest pairs algorithm*. Available from: <http://www.brunel.ac.uk/~castjjg/java/shortest_path/shortest_path.html> (Visited: 9/12/2003).

Griffith, B. C., Small, H., Stonehill, J. A., and Dey, S. (1974). The structure of scientific literature, II: toward a macro and microstructure for science. *Science Studies*, 4, 339-365.

Grupo SCImago. (2002). *Imago Scientae - Science Visualization*. Available from: <http://www.scimago.es> (Visited: 31/3/2005).

Guerrero Bote, V.P. (1997). Redes neuronales aplicadas a las técnicas de recuperación documental. Tesis Doctoral. Granada: Universidad, Departamento de Biblioteconomía.

Guerrero Bote, V.P., Moya-Anegón, F.d., and Herrero Solana, V. (2002a) Automatic extraction of relationships between terms by mean of kohonen's algoritm. *Library & Information Sciences Research*, 24, 235-250.

Guerrero Bote, V.P., Moya-Anegón, F.d., and Herrero Solana, V. (2002b) Document organization using Kohonen's algorithm. *Information Processing & Management*, 38, 79-89.

Herrero, R. (1999) La terminología del análisis de redes: problemas de definición y traducción. *Revista Política y Sociedad*, 33, 11 p.

Herrero Solana, V. (1999). Modelos de representación visual de la información bibliográfica: aproximaciones multivariante y conexionistas. Tesis Doctoral. Granada: Universidad, Departamento de Biblioteconomía y Documentación.

Herrero Solana, V. and Moya-Anegón, F.d. (2001). Bibliographic displays of web-based OPACs: multivariate analysis applied to Latin-American catalogues. *Libri*, 51, 75-85.

Hjørland, B. and Albrechtsen, H. (1995). Toward a new horizon in information science: domain analysis. *Journal of the American Society for Information Science (JASIS)*, 46, 400-425.

Hjørland, B. (2002). Domain analysis in information science: eleven approaches-traditional as well as innovative. *Journal of Documentation (JDOC)*, 58, 422-462.

Ingwersen, P. (2001). Cognitive perspective of representation. In: V Congreso Isko-España. La representación y organización del conocimiento: metodologías, modelos y aplicaciones, (32-41), Madrid: Ana Extremeño Placer.

Ingwersen, P. and Larsen, B. (2001). Mapping national research profiles in social science disciplines. *Journal of Documentation*, 57, 715-740.

Jackson, D. (2002). SVG on the rise. [On-line]. http://www.oreillynet.com/pub/a/javascript/2002/06/06/svg_future.html [Visited: 11/09/2003].

Jarneving, B. (2001). The cognitive structure of current cardiovascular research. *Scientometrics* 50, 365-389.

Jiménez Contreras, E., Moya-Anegón, F.d., and Delgado López-Cózar, E. (2003). The evolution of research activity in Spain. The impact of the National Commission for the Evaluation of Research Activity (CNEAI). *Research Policy,* 32, 123-142.

Kamada, T. and Kawai, S. (1989). An algorithm for drawing general undirected graphs. *Information Processing Letters,* 31, 7-15.

Kaski, S., Honkela, T., Lagus, K., and Kohonen , T. (1998). Websom: self - organizing maps of document collection. *Neurocomputing,* 21, 101-117.

Klovdhal, A.S. (1981). A note of images of social networks. *Social Networks,* 3, 197-214.

Kohonen , T. (1985). The self-organizing map. *Proceedings of the IEEE,* 73, 1551-1558.

Kohonen , T. (1997). Self-organizing maps. Berlin [etc.]: Springer-Verlag.

Krempel, L. (1999). Visualizing Networks with Spring Embedders: Two-mode and Valued Graphs. International Sunbelt Social Network Conference.

Kruskal, J.B. and Wish, M. (1978). Multidimensional Scaling. London: Sage.

Kyvik, S. (2003). Changing Trends in Publishing Behaviour among University Faculty, 1980-2000. Scientometrics, 58 (1), 35-48.

Lewis-Beck, M.S. (1994). Factor analysis and related techniques. London: Sage.

Liberman, S. and Wolf, K. B. (1997). The Flow of Knowledge: Scientific Contacts in Formal Meetings. *Social Networks,* 19, (3), 271-283.

Lin, X. (1997). Map displays for information retrieval. *Journal of the American Society for Information Science (JASIS),* 48, 40-54.

Lin, X., Soergel, D., and Marchionini, G. (1991). A self-organizing semantic map for information retrieval. In: Proceedings of the Fourteenth Annual International ACM/SIGIR Conference on Research and Development in Information Retrieval, (262-269), Chicago.

Lin, X., White, H. D., and Buzydlowski, J. (2003). Real-time author co-citation mapping for online searching. *Information Processing & Management,* 689-706.

Macromedia. (2005). *Flash MX 2004.*

Marion, L.S. and McCain, K.W. (2001). Contrasting views of software engineering journals: Author cocitation choices and indexer vocabulary assignments. *Journal of the American Society for Information Science and Technology (JASIST),* 52, 297-308.

Martínez Arias, R. (1999). El análisis multivariante en la investigación científica. Madrid: La Muralla.

McCain, K.W. (1990). Mapping authors in intellectual space: a technical overview. *Journal of the American Society for Information Science (JASIS),* 41, 433-443.

McCain, K.W. (1991a). Core journal networks and cocitation maps: new bibliometrics tools for serial research and management. *Library Quarterly,* 61, 311-336.

McCain, K.W. (1991b). Mapping economics throgh the journal literature: an experiment in journal cocitationanalysis. *Journal of the American Society for Information Science (JASIS)*, 42, 291-296.

McCormick, B.H., DeFanti, T.A., and Brown, M.D. (1987). Visualization in Scientific Computing. *Computers Graphic*, 21, 17-32.

Merton, R. K. (2000). On the Garfield input to the sociology of science: a retrospective collage. In B. Cronin and H. B. Atkins (Eds.). The web of knowledge: a festschrift in honor of Eugene Garfield. New Jersey: Information Today.

Molina, J.L. (2001). El análisis de redes sociales: una introducción. Barcelona: Bellaterra.

Molina, J. L., Muñoz, J., and Losego, P. (2000). Red y realidad: aproximación al análisis de redes científicas. In: VI Congreso Nacional de Psicología Social, (21 p.).

Moreno, J.L. (1934). Who shall survive? Washington, DC: Nervous and Mental Disease Publishing Company.

Moreno, J.L. (1953). Who shall survive? New York: Beacon House Inc.

Morris, T.A. (1998). The structure of medical informatics journal literature. *Journal of the American Society for Information Science and Technology (JASIST)*, 5, 448-466.

Moya-Anegón, F.d., Vargas-Quesada, B., Chinchilla-Rodríguez, Z., Herrero-Solana, V., Corera-Álvarez, E., and Munoz-Fernández, F.J. (2004). A new technique for building maps of large scientific domains based on the cocitation of classes and categories. Scientometrics, 61(1), 129-145. 2004.

Moya-Anegón, F.d., Vargas-Quesada, B., Chinchilla-Rodríguez, Z., Herrero-Solana, V., Corera-Álvarez, E., and Munoz-Fernández, F.J. (2005). Domain analysis and information retrieval through the construction of heliocentric maps based on ISI-JCR category cocitation. *Information Processing & Management*, 41, 1520-1533.

Moya-Anegón, F.d. and Herrero Solana, V. (1999). Investigaciones en curso sobre interfaces gráficos en dos y tres dimensiones para el acceso a la información electrónica. *Cuadernos de Documentación Multimedia*, 8.

Moya-Anegón, F.d., Herrero Solana, V., Vargas-Quesada, B., Chinchilla-Rodriguez, Z., Corera-Alvarez, E., Muńoz Francisco, Olvera-Lobo, D., Fernández-Molina, J.C., García-Santiago, D., Guerrero Bote, V.P., Faba-Pérez, C., López-Pujalte, C., Reyes-Barragán, M., and Zapico-Alonso, F. (2004). Atlas de la Ciencia Española: propuesta de un sistema de información científica. *Revista Española de Documentación Científica*, 27, 11-29.

Moya-Anegón, F.d. and Herrero-Solana, V. (2001). Análisis de dominio de la investigación bibliotecológica mexicana. *Información, cultura y sociedad* 5.

Moya-Anegón, F. d., Herrero-Solana, V., and Guerrero Bote, V. P. (1998). La aplicación de redes neuronales artificiales (RNA) a la recuperación de la información. In Baró i Queralt, J. and Cid Leal, P. Eds. Anuari SOCADI de Documentació i informació, (147-164), Barcelona: Societat Catalana de Documentació i Informació.

Moya-Anegón, F.d., Jiménez Contreras, E., and Moneda Carrochano, M.d.l. (1998) Research fronts in library and information science in Spain (1985-1994). *Scientometrics,* 42, 229-246.

Moya-Anegón, F. d., Moscoso, P., Olmeda, C. O.-R. V., Herrero, V., and Guerrero, V. (1999). Neurolsoc: un modelo de red neuronal para la representación del conocimiento. In: IV Congreso Isko-España EOCONSID'99, (151-156), Granada: Maria José López-Huertas, Juan Carlos Fernández-Molina.

Orwig, R.E., Chen, H., and Nunamaker, J.F.Jr. (1997). A graphical, self-organizing approach to classifying electroninc meeting output. *Journal of the American Society for Information Science (JASIS),* 48, 157-170.

Owen, G. S. (1999). *Framework of a visualization system.* Available from: <http://www.siggraph.org/education/materials/HyperVis/abs_con1/main.htm> (Visited: 2/8/2005).

Persson, O. (1994). The intellectual base and research fronts of JASIS: 1986-1990. *Journal of the American Society for Information Science (JASIS),* 45, 31-38.

Polanco, X., Francois, C., and Keim J. P. (1998). Artificial neural network technology for the classification and cartography of scientific and technical information. *Scientometrics,* 41, 69-82.

Price, D.d.S. (1965). Networks of scientific papers. *Science,* 149, 510-515.

Regueiro, C.; Barro, S.; Sánchez, E.; y Fernández-Delgado, M. (1995). Modelos básicos de redes neuronales artificiales. In S. Barro and J. E. Mira (Eds.), Computación neuronal Santiago de Compostela: Universidad: Servicio de Publicaciones e Intercambio Científico.

Rice, R.E., Borgman, C.L., and Reeves, B. (1988). Citation networks of communication journals, 1977-1985. *Human communication research.* 15, 256-283.

Rodriguez, J.A. (1995). Análisis estructural y de redes. Madrid: CIS.

Salton, G., Allan, J., and Buckley, C. (1994). Automatic structuring and retrieval of large text file. *Communications of the ACM,* 37, 97-108.

Salton, G. and Bergmark, D. (1979). A citation study of computer science literature. *Professional Communication, IEEE Transaction,* PC-22, 146-158.

Salton, G. and McGill, M.J. (1983). Introduction to modern information retrieval. New York: McGraw-Hill.

Sanz E., Aragón I., and Méndez, A. (1995). The function of journals in disseminating applied science . *Journal of Information Science,* 21, 319-323.

Schvaneveldt, R.W. (1990). Pathfinder Associative Networks. Norwood, NJ: Ablex.

Scott, J. (1992). Social network analysis: a handbook. London: Sage.

Small, H. (1973). Co-citation in the scientific literature: a new measure of the relationship between two documents. *Journal of the American Society for Information Science (JASIS),* 24, 265-269.

Small, H. (1997). Update on science mapping: creating large document spaces. *Scientometrics,* 38, 275-293.

Small, H. (1999). Visualizing science by citation mapping. *Journal of the American Society for Information Science (JASIS),* 50, (9), 799-813.

Small, H. (2000). Charting pathways through science: exploring Garfield's vision of a unified index to science. In B. Cronin and H. B. Atkins (Eds.), The web of knowledge: a festschrift in honor of Eugene Garfield Medford. New Jersey: Information Today.

Small, H. and Garfield, E. (1985). The geography of science: disciplinary and national mappings. *Journal of Information Science,* 11, 147-159.

Small, H. and Griffith, B. C. (1974). The structure of scientific literature, I: identifying and graphing specialyties. *Science Studies,* 4, 17-40.

Sugiyama, K., Tagawa, S., and Toda, M. (1981). Methods for visual understanding of hierachical system structures. *IEEE Transaction on Systems Man and Cibernetics,* SMC-11, 109-125.

Tamassia, R., Batista, G., and Batini, C. (1988). Automatic graph drawing and readability of diagrams. *IEEE Transaction on Systems Man and Cibernetics* SMC-18, 61-79.

The Thomson Corporation. (2005a). *ISI Journal Citation Reports.* Available from: <http://go5.isiknowledge.com/portal.cgi> (Visited: 3/9/2005a).

The Thomson Corporation. (2005b). *ISI Web of Knowledge.* Available from: <http://go5.isiknowledge.com/portal.cgi> (Visited: 3/9/2005c).

Thurstone, L.L. (1931). Multiple factor analysis. *Psycological review,* 38, 406-427.

Tufte, E.R. (1994a). Envisioning information. Cheshire: Graphics Press.

Tyron, R.C. (1939). Cluster analysis. New York: Mc-Graw-Hill.

W3C. (2001). *Scalable Vector Graphics (SVG) 1.0 Especificación.* Available from: <http://www.w3.org/TR/SVG/> (Visited: 15/9/2003).

W3C. (2003a). *Extensible Markup Language (XML).* Available from: <http://www.w3.org/XML/> (Visited: 9/10/2003a).

W3C. (2003b). *HiperText Markp Language (HTML).* Available from: <http://www.w3.org/MarkUp/> (Visited: 9/12/2003b).

W3C. (2003c). *Scalable Vector Graphics (SVG).* Available from: <http://www.w3.org/Graphics/SVG/> (Visited: 9/10/2003c).

W3C. (2004). *Scalable Vector Graphics (SVG) 1.1 Specification.* Available from: <http://www.w3.org/TR/SVG11/> (Visited: 9/10/2004).

Wasserman, S. and Faust, K. (1998). Social network analysis: methods and applications. Cambridge: Cambridge University Press.

Watts, D.J. and Strogatz, S.J. (1998). Collective dynamics of small-world. networks. *Nature,* 393, 440-442.

Welman, B. (1988). Structural analysis: from method and metaphor to theory and substance. In B. Welman and S. D. Berkowitz (Eds.), Social Structures a Network Approach. Cambridge: University Press.

White, H. D. (2000). Toward ego-centered citation analysis. In B. Cronin and H. B. Atkins (Eds.). The web of knowledge: a festschrift in honor of Eugene Garfield (pp. 475-498). New Jersey: Information Today.

White, H.D. (2001). Author-centered bibliometrics through CAMEOs: characterizations automatically made and edited online. *Scientometrics,* 51, 607-637.

White, H. D. (2003). Pathfinder networks and author cocitation analysis: a remapping of paradigmatic information scientist. *Journal of the American Society for Information Science and Technology (JASIST),* 54, (5), 423-434.

White, H. D., Buzydlowski, J., and Lin, X. (2000). Co-cited author maps as inter-faces to digital libraries: designing Pathfinder Networks in the humanities. In: IEEE International Conference on information visualization, (25-30), London.

White, H.D. and Griffith, B.C. (1981a). Author cocitation: a literature measure of intellectual structure. *Journal of the American Society for Information Science (JASIS)*, 32, 163-172.

White, H. D. y Griffith, B. C. (1981b). A cocitation map of authors in judgment and decision research. In B.F. Anderson [et al.] (Ed.), Concepts in judgments and decision research: definition sources interrelationships, and comments. New York: Praeger.

White, H.D. and Griffith, B.C. (1982). Authors as markers of intellectual space: cocitation in studies of science, technology and society. *Journal of Documentation*, 38, 255-272.

White, H. D., Lin, X., and McCain, K. W. (1998). Two modes of automated domain analysis: multidimensional scaling vs. Kohonen feature mapping of information science authors. In: Proceedings of the Fifth International ISKO Conference, (57-61), Würzberg: Ergon Verlag.

White, H.D. and McCain, K.W. (1997). Visualization of literature. *Annual Review of Information Systems and Technology (ARIST)*, 32, 99-168.

White, H.D. and McCain, K.W. (1998). Visualizing a discipline: an author co-citation analysis of information science, 1972-1995. *Journal of the American Society for Information Science (JASIS)* 49, 327-355.

Printing: Krips bv, Meppel
Binding: Stürtz, Würzburg